Oil Spill Dispersants

Mechanisms of Action and Laboratory Tests

John R. Clayton, Jr.
Science Applications International Corporation
San Diego, California

James R. Payne
Sound Environmental Services, Inc.
Carlsbad, California

John S. Farlow
Releases Control Branch, Risk Reduction Engineering Laboratory
U.S. EPA
Edison, New Jersey

Choudhry Sarwar
EPA Technical Project Monitor

CRC Press
Taylor & Francis Group
Boca Raton London New York

CRC Press is an imprint of the
Taylor & Francis Group, an **informa** business

First published 1993 by C. K. SMOLEY

Published 2019 by CRC Press
Taylor & Francis Group
6000 Broken Sound Parkway NW, Suite 300
Boca Raton, FL 33487-2742

© 1993 by Taylor & Francis Group, LLC
CRC Press is an imprint of Taylor & Francis Group, an Informa business

First issued in paperback 2019

No claim to original U.S. Government works

ISBN-13: 978-0-367-45006-9 (pbk)
ISBN-13: 978-0-87371-946-9 (hbk)

Visit the Taylor & Francis Web site at
http://www.taylorandfrancis.com

and the CRC Press Web site at
http://www.crcpress.com

Library of Congress Cataloging-in-Publication Data

Oil spill dispersants : mechanisms of action and laboratory tests /
  John R. Clayton, Jr. . . . [et al.].
      p. cm.
    Includes bibliographical references (p.) and index.
    ISBN 0-87371-946-8
    1. Oil spills—Environmental aspects. 2. Dispersing agents-
-Testing. I. Clayton, John R., 1947–
TD427.P4O3864   1992
628.5′2—dc20   92-35044
                                                                  CIP

Acknowledgment

The authors would like to thank Mr. Choudhry Sarwar, the U.S. Environmental Protection Agency Technical Project Monitor, for his continued help and support throughout the program in which the information in this book was compiled. As noted in the Appendix, the authors also are indebted to a number of experts on chemical dispersants for their input and review of preliminary drafts of the EPA-sponsored report that was central to the development of this book. Finally, the authors would like to express their gratitude to Mr. Siu-Fai Tsang, Ms. Victoria Frank, Dr. Paul Marsden, and Mr. John Harrington of Science Applications International Corporation for their support and efforts in conducting laboratory studies for dispersant performance with a variety of testing procedures. Some results from the latter studies are included in this book.

Notice

FOREWORD

Today's rapidly developing and changing technologies and industrial products and practices frequently carry with them the increased generation of materials that, if improperly dealt with, can threaten both public health and the environment. The U.S. Environmental Protection Agency is charged by Congress with protecting the Nation's land, air, and water resources. Under a mandate of national environmental laws, the Agency strives to formulate and implement actions leading to a compatible balance between human activities and the ability of natural systems to support and nurture life. These laws direct the EPA to perform research to define our environmental problems, measure the impacts, and search for solutions.

The Risk Reduction Engineering Laboratory is responsible for planning, implementing, and managing research, development, and demonstration programs to provide an authoritative, defensible engineering basis in support of the policies, programs, and regulations of the EPA with respect to drinking water, wastewater, pesticides, toxic substances, solid and hazardous wastes, and Superfund-related activities. This publication is one of the products of that research and provides a vital communication link between the researcher and the user community.

The purpose of this book is to summarize the most current information from the available literature for the mechanism of action of chemical dispersants (or how they work) for oil spills, variables that affect dispersant performance, evaluations of a variety of laboratory tests designed to assess the performance of dispersant agents, and a brief consideration of relationships between laboratory methods and field situations. In part, the book updates information on the preceding topics from earlier documents (e.g., NRC, 1989; Nichols and Parker, 1985; Mackay et al., 1984; Meeks, 1981; Rewick et al., 1981). Considerations are given to strengths and limitations of specific laboratory tests, including brief discussions of the applicability of their results for estimating dispersant performance in field trials or conditions encountered during real spill events. Finally, a modest attempt is made at providing recommendations for improvements in future laboratory testing.

This review summarizes work from both the peer-reviewed scientific literature as well as scientific reports not submitted for formal publication. For expediency, liberal use is made of information presented in several excellent reviews that consider mechanisms of dispersant action and laboratory testing for dispersant performance (e.g., NRC, 1989; Nichols and Parker, 1985; Mackay et al., 1984; Rewick et al., 1984; Meeks, 1981; Rewick et al., 1981; and references therein). While much information exists in the published and unpublished literature on relative toxicities of dispersant and dispersant-oil mixtures, this topic is beyond the scope of this review.

ABSTRACT

Discussions are presented for (1) the mechanism of action of chemical dispersants for oil spills, (2) factors affecting performance of dispersants and its measurement, (3) some common laboratory methods that have been used to test dispersant performance, (4) a brief summary of dispersant applications and their performance in field trials and spills-of-opportunity, and (5) recommendations for future laboratory studies. The discussion of laboratory methods for performance testing presents information regarding the approach used for general laboratory tests, detailed information for a number of the more commonly used tests, and similarities and differences among tests. Differences among tests are important because they may be responsible for not only significant differences in results between laboratory testing methods but also poor correlations between laboratory results and data from field tests. Four general types of laboratory testing methods are considered: (a) tank tests, (b) shake/flask tests, (c) interfacial surface tension tests, and (d) flume tests. For each test considered, descriptions are presented for the laboratory apparatus required, brief summaries of the testing procedures, differences among methods, and considerations of how a particular test design might affect results. With the understanding that the purpose of this book centers on laboratory testing, a brief discussion is presented of field trials that have involved dispersant applications. Information is presented for general approaches used in studies, limitations encountered in such efforts, and how realistic it is to compare laboratory results with field data. Brief descriptions are also presented for a number of rapid field tests for estimating dispersant performance. Limitations inherent to measurements obtained with the latter tests are discussed.

The report that is the basis for this book was submitted in partial fulfillment of EPA Contract No. 68-C8-0062 by Science Applications International Corporation under the sponsorship of the U.S. Environmental Protection Agency.

John R. Clayton, Jr. is a Senior Scientist in the Environmental Sciences Division, Science Applications International Corporation, San Diego, California. He received a Ph.D. in Fisheries from the University of Washington-Seattle in 1985, a M.S. in Oceanography from the University of Washington-Seattle in 1976, and a B.S. with honors in Zoology from the University of California-Los Angeles in 1970. His Ph.D. research focused on the biochemical metabolism of nitrogen in marine phytoplankton. The studies included methods development for measurements of a variety of biochemical parameters (e.g., *in vitro* assays for the enzymes glutamate synthase and glutamine synthetase as well as improved quantification techniques for total proteins, total free amino acids, RNA, and DNA), utilization and adaptation of autoanalyzer techniques for chemical measurements of inorganic nutrients and nucleic acids, and evaluations of effects of nitrogen deprivation on cellular distributions of nitrogen and enzyme activities in phytoplankton. His M.S. research focused on analysis of samples and evaluation of environmental factors contributing to concentrations of PCBs in natural populations of marine zooplankton. Between his M.S. and Ph.D. programs, research at the University of California Bodega Marine Laboratory included analysis and methods development for measurements of oil hydrocarbons and chlorinated pesticides in seawater, suspended sediment, tissue, and waterfowl samples.

As a Senior Scientist at Science Applications International Corporation, he has been responsible for technical design, implementation, and project management for studies dealing with evaluation, monitoring, and remediation of chemical pollutants in coastal and estuarine regions. Examples of studies include updating state-of-the-art information on mechanisms of action as well as testing and evaluation of laboratory procedures for chemical dispersants and shoreline-cleaning-agents for treating oil spills; evaluation of effects of chemical dispersants on the behavior and fate of oil spilled in a simulated freshwater streambed; measurement of petroleum hydrocarbons in sediment and water samples for the U.S. EPA Bioremediation study in Alaska after the EXXON VALDEZ oil spill; measurement of oil and polar degradation products in natural samples of sediment, seawater, and suspended particulate material following the EXXON VALDEZ spill; evaluation of interactions of crude and refined oil products, suspended sediments, and ice in turbulent and non-turbulent water systems in subarctic and arctic environments; and analysis of polynuclear aromatic hydrocarbons, chlorinated pesticides and PCBs in natural marine samples of sediments and bivalve tissues.

James R. Payne is a Senior Vice President and Director of Research for Sound Environmental Services, Inc. in Carlsbad, California. He received his Ph.D. in chemistry from the University of Wisconsin-Madison in 1974 and a B.A. with honors from California State University-Fullerton in 1969. He was a National Institutes of Health Predoctoral Fellow at Wisconsin and focused his research on apoenzyme/coenzyme interactions and the organic synthesis of coenzyme analogs, including carbon-13 enriched vitamin B-6, for carbon-13 nuclear magnetic resonance studies of enzyme/coenzyme active site complexes. After graduate school, he received a Woods Hole Oceanographic Institution Postdoctoral Scholarship, where he undertook research on marine humic acids, incorporation of petroleum hydrocarbons into marine shellfish, and the persistence and metabolism of PCBs in the water column of the North Atlantic.

Later research at the University of California Bodega Marine Laboratory centered on development of laboratory methods and large-volume seawater sampling systems for detection of trace-level organics, particularly petroleum hydrocarbons, in seawater and tissues of selected marine species. As a Senior Chemist and Assistant Vice President at Science Applications International Corporation, he was an invited scientist on the NOAA ship RESEARCHER, investigating the IXTOC I oil well blowout in the Gulf of Mexico in 1979, and Principal Investigator on over 11 years of experimental and modeling efforts on oil weathering behavior in subarctic and arctic marine environments. These projects culminated in extensive study of the 1989 EXXON VALDEZ oil spill in Prince William Sound, Alaska. Dr. Payne also was Chief Scientist on Spill of Opportunity Dispersant Trials at the PAC BARONESS oil spill off Point Conception, California in 1987 and the MEGA BORG oil spill off Galveston, Texas in 1990. In addition, he was a member of the National Academy of Sciences/National Research Council Committee on Effectiveness of Oil Spill Dispersants leading to the publication of *Using Oil Spill Dispersants on the Sea* (NRC, 1989).

John S. (Jack) Farlow is Chief of the Releases Control Branch of the U.S. Environmental Protection Agency's Risk Reduction Engineering Laboratory. He manages a group of scientists and engineers who develop and improve technology relating to oil spill, leaking underground storage tank, and abandoned industrial waste site (Superfund) cleanups.

Mr. Farlow received a B.A. in geology from Harvard University in 1957 and a M.A. in Physical Oceanography from the Johns Hopkins University in 1960. Following completion of his M.A., he worked on water quality management studies in the Great Lakes and in northeastern estuarine systems. In 1972 he joined the U.S. EPA's oil spill engineering research group, overseeing the construction and operation of the OHMSETT wave/tow-tank facility as well as a variety of research projects. In 1978 he became Chief of the Oil Spills Staff (managing all EPA oil spill engineering and chemical countermeasures research). He has served as Branch Chief since 1988. Mr. Farlow also has served on the International Oil Spill Conference Steering Committee since 1982.

TABLE OF CONTENTS

TABLE OF CONTENTS (continued)

List of Tables

List of Figures

SECTION 1

INTRODUCTION

In the event of unintentional releases of oil into coastal waters, oil from slicks can have deleterious impacts to biota in exposed ecosystems. Effects will depend in large part on the ultimate location of the oil as well as its chemical composition at the time of interaction with the biota (e.g., NRC, 1985; Capuzzo, 1987; Spies, 1987). Oil that remains on the water's surface as a slick can produce major impacts to organisms that associate with the air-water interface (e.g., sea birds and air-breathing animals such as sea otters, pinnipeds, and cetaceans). Deleterious impacts can derive from both physical and chemical effects of the oil. For example, oil can coat the plumage and fur of birds and fur-bearing marine mammals, respectively (e.g., Hunt, 1987; Geraci and St. Aubin, 1987). Ingestion of oil as well as loss of thermal insulation properties from the oiled feathers or fur can produce severe consequences to the organisms. Unintentional inhalation of oil by aquatic, air-breathing organisms can produce respiratory complications. At the same time, dispersion of oil as droplets from surface slicks into the water column (e.g., by either natural or human-mediated processes) will lessen impacts to organisms at the air-water interface, but enhance exposure to biota in the water column if the oil is not sufficiently diluted. Should the oil sink (e.g., following interaction with sedimentary or particulate matter to produce densities in the oil matrix greater than that of seawater), benthic communities also can be affected.

If treatment of an oil slick on water is determined to be necessary or justified (i.e., as opposed to doing nothing), then a number of remediation approaches are possible. Four cleanup strategies that frequently receive consideration include: (1) mechanical cleanup or recovery, (2) burning, (3) bioremediation, and (4) treatment with chemical dispersants. Brief discussions of each of these options is presented below, although the major focus of this book is to address chemical dispersants and evaluation of their performance in the laboratory in greater detail.

Mechanical cleanup and removal of oil from a surface slick on water is a common and highly visible corrective-action option that has been used for mitigation of oil spills. Such activities can include corralling of slicks (e.g., by booms and tow vessels), skimming of slicks from water surfaces by specially designed apparatus and vessels, utilization of equipment constructed of synthetic polymers that enhance collection of oil and minimize incorporation of water that forms intractable water-in-oil emulsions, and application of high-pressure water (hot or cold) jets to remove oil from surfaces (e.g., stranded oil on shorelines). All of these options were used in efforts to recover oil following the EXXON VALDEZ spill incident. While the procedures met with mixed success in Prince William Sound and along the outer coast of south-central Alaska (depending on the specific situation in which they were used), a major limitation inherent to all mechanical cleanup approaches involves the necessity for disposing of collected oil wastes and materials at the conclusion of the corrective action.

Burning of an oil slick on water is attractive in remote areas where access to a spill is difficult. Burning requires a minimum of equipment to perform the remediation action. Furthermore, all but a few percent of the oil is normally converted to gasified products during the combustion, leaving only a few percent of the original spill volume as residue at the conclusion of a burn. Consequently, burning minimizes the need for physical collection, storage, and transport of residual, recoverable product. However, successful burning of oil on water requires that the thickness of the slick be sufficient to support efficient burning of the oil. An example of a successful test burn of spilled oil in Prince William Sound during the second day after the EXXON VALDEZ incident is reported by Allen (1991). The oil for the burn was collected on the water's surface by towing Fire Boom behind two fishing vessels. The quantity of oil collected within the towed-boom arrangement was estimated to be o᷃ the order of 10,000 gallons (38,000 L). The average thickness of the slick was estimated at 0.3-1 mm, with patches as thick as 2-3 mm. Ignition was accomplished by introducing a lighted bag of gelled gasoline (approximately 0.5 L of fuel and a handful of Sure Fire gelling mix) into the slick. The entire burn period (including ignition) lasted approximately 75 minutes, with the most intense burning occurring during a 45-minute period. The thickness and location of burning oil within the boom was controlled within certain limits by adjusting the speed of the towing vessels. Average oil thickness in the burning area was estimated to be on the order of 13-18 cm (5-7 inches) based on values for the towing speed of the vessels, the estimated quantity of total oil contained in the boom, and the size of the burning area. At the conclusion of the burn, estimates indicated that only approximately 300 gallons (1136 L) of burn residue (which had a taffy-like consistency unlike the parent oil) remained on the water's surface. Consequently, approximately 97% of the oil was estimated to have been removed during the burn event.

While burn studies like that described above appear to be quite successful, efficient burning of the oil does require that slick thicknesses be considerably greater than that normally occurring in slicks that experience natural spreading on the water's surface. Although oil slicks at sea are generally non-uniform in thickness and distribution, many spills of widely varying size appear to reach an average thickness of approximately 0.1 mm (NRC, 1989). In addition to requiring mechanical collection or corralling of the oil to achieve sufficient thickness for successful burning, additional concerns for utilization of burning include overall safety of the burning process itself and the impact of the smoke (i.e., soot) on public and environmental health. Studies have indicated that additions of certain chemicals (e.g., butylferrocene and pentylferrocene) to oils can increase burning efficiency and reduce emission quantities of smoke and soot (Mitchell and Janssen, 1991). However, mechanisms for introduction of such burn-enhancers to an oil slick remain to be developed.

Bioremediation is another treatment option that has received consideration for treatment of oil slicks. In bioremediation, oil is biodegraded by natural processes that are primarily mediated by microorganisms (e.g., bacteria, fungi, unicellular algae, and protozoa) following release of the oil onto/into an aquatic environment. Enzymes synthesized by the microorganisms catalyze the oxidation of oil constituents to intermediate metabolites, which can be either incorporated into biomass or ultimately metabolized (by the starting or subsequent microorganisms) to carbon dioxide and water. As such, oil is removed and a final cleanup is not required. Biodegradation can occur in situations where an oil-water interface, nitrogen, phosphorus, oxygen, and a suitable assemblage of microorganisms exists. The microorganisms are present in the aqueous phase and metabolize oil at the oil-water interface as a carbon source if sufficient nutrients (e.g., nitrogen and phosphorus) and oxygen are present for the metabolic reactions to occur. Nutrients and/or oxygen frequently are considered to be limiting for biodegradation processes because of the large excess of carbon present due to the oil. Additional factors that can influence rates of biodegradation of oil include temperature, the chemical composition of the oil (which changes over time due to natural weathering processes), and the total surface area of the oil-water interface.

While biodegradation is a natural process, bioremediation is defined as any technique that enhances the natural rate of biodegradation. Specifically, bioremediation normally involves addition(s) of either nutrient supplements (e.g., nitrogen and phosphorus) or "oil-consuming" microbial inocula to an oil-contaminated system. Addition of only a nutrient supplement is intended to enhance biodegradation processes in indigenous populations of microorganisms. While not conclusive, information in the literature suggests that nutrient supplements can lead to enhanced rates of oil degradation, although such instances normally occur on oiled shorelines as opposed to surface slicks on open water (e.g., the EXXON VALDEZ oil spill; Chianelli et al., 1991; Pritchard and Costa, 1991; Tabak et al., 1991). Bioremediation products based on microbial inocula usually incorporate a mixture of "oil-consuming" microbes (i.e., mixtures of oil-degrading microbes that have been collected from assorted parts of the world) and nutrients. The limited information available does not appear to indicate that microbial-inocula products substantially enhance rates of biodegradation of oil in natural systems beyond that due to the nutrient enhancement of indigenous microbial populations alone (e.g., Owen, 1991; Venosa et al., 1991).

In aquatic systems, bioremediation agents can be applied to an oil slick on water. However, successful removal of a slick by bioremediation agents in such situations appear to suffer from severe limitations including: (1) successful application of agents onto/into an oil slick can be problematic, (2) applied agents are likely to be rapidly lost from the slick to the adjacent water phase, and (3) a successfully treated slick may well go ashore in coastal situations before the bioremediation agent has time to substantially enhance the biodegradation of the oil.

The final strategy identified above for remediating oil spills on water involves treatment with chemical dispersant agents. The essential components in dispersant formulations are surfactant molecules. These are compounds containing both oil-compatible (i.e., lipophilic or hydrophobic) and water-compatible (i.e., hydrophilic) groups. Following successful application of a chemical dispersant formulation to an oil slick on water, the surfactant molecules will reside at oil-water interfaces and reduce the oil-water interfacial surface tension. In the presence of minimal mixing energy (e.g., provided by wave or wind action), this lowering of the oil-water interfacial surface tension will result in dispersion of the oil as small droplets into the underlying water column. Such dispersion is intended to lead to not only overall dilution of the oil in the water but also increased oil-water interfacial surface area, which should favor microbial degradation of the oil at the oil-water interfaces. The purpose is removal of oil from the water's surface, followed by dilution and degradation of the oil to non-problematic concentrations in an underlying water column. If successful, no recovery of final oil residues from either the water's surface or the water column should be required. Furthermore, the likelihood of an oil slick stranding on a shoreline is minimized.

The primary focus of this book is to consider the mechanism of action of chemical dispersant agents as well as the testing or evaluation of their performance in the laboratory. As described in Cunningham et al. (1991), the use of chemical dispersants to respond to oil spills in U.S. waters is governed by Subpart J of the National Contingency Plan (NCP; 40 CFR 300.900). If criteria for application of chemical dispersants in a spill situation are satisfied, a federal on-scene coordinator (OSC) at an oil spill has the option to use chemical dispersants that are listed on the NCP Product Schedule. For a product to be listed on the Product Schedule, the manufacturer must first submit technical data on the product to the U.S. Environmental Protection Agency (EPA). For a dispersant agent, the required technical data at the present time include laboratory results for a standardized dispersant effectiveness test and a standardized aquatic toxicity test. While dispersant effectiveness and aquatic toxicity tests are currently defined, no criteria for acceptance levels (i.e., "pass/fail") presently exist for either test. Furthermore, the current dispersant effectiveness test (the Revised Standard EPA test; 40 CFR Part 300; EPA, 1984) involves not only a relatively complex testing apparatus that is not readily amenable to rapid turn-around of laboratory test results but also generates relatively large volumes of oily waste water (130 L). In addition to the detailed discussion of the mechanism of action of chemical dispersant agents, this book also describes a variety of laboratory testing procedures (including the Revised Standard EPA test) that have the potential to be used for evaluating the performance of commercial dispersant agents. Such procedures can be used to evaluate the performance of chemical agents for dispersing oil slicks on water in relatively controlled laboratory environments. The results also might be used for purposes such as identifying agents that could be useful for dispersing oil slicks in the natural environment.

SECTION 2

GENERAL MECHANISM OF ACTION OF CHEMICAL DISPERSANTS

The following information regarding the mechanism of action of chemical dispersants is summarized in large part from the work of a number of investigators (e.g., Bobra, 1991; Bridie et al., 1980; Brochu et al., 1986/87; Canevari, 1978, 1982, 1984, 1987, 1991; Farn, 1983; NRC, 1989). Briefly, chemical dispersants are intended to promote the break-up or dispersion of an oil slick into small droplets that distribute into a water column. The small oil droplets should not recombine or coalesce to reform surface slicks. Ideally, the dispersed oil droplets will be subjected to dilution to non-problematic concentrations in the water column as well as enhanced microbial degradation with the increase in the amount of oil-water interfaces.

A typical commercial chemical dispersant is a mixture of three types of chemicals: surface active agents (i.e., surfactants), solvents, and additives. Solvents are primarily present to promote the dissolution of surfactants and additives into a homogeneous dispersant mixture. Additives may be present for a number of purposes such as increasing the biodegradability of dispersed oil mixtures, improving the dissolution of the dispersant into an oil slick, and increasing the long-term stability of the dispersion. For the actual dispersion process, however, the most important components in the dispersant mixture are the surfactant molecules. These are compounds containing both oil-compatible (i.e., lipophilic or hydrophobic) and water-compatible (i.e., hydrophilic) groups. Because of this amphiphatic nature (i.e., opposing solubility tendencies), a surfactant molecule will reside at oil-water interfaces as shown in Figure 1. By the indicated orientation at the interface (i.e., hydrophobic and hydrophilic groups positioned toward the oil and water phases, respectively), the surfactant will reduce the oil-water interfacial surface tension.

The relationship between oil-water interfacial surface tension and the concentration of chemical dispersant (or surfactant molecules) is shown in Figure 2, as defined by G. Lindblom (personal communication). As the concentration of individual surfactant molecules residing at the oil-water interface increases, the oil-water interfacial surface tension becomes progressively lower in proportion to the increase in the number of surfactant molecules. The Critical Micelle Concentration (CMC) is the concentration of dispersant at which surfactant molecules form a uniform monolayer at the oil-water interface. Further increase in the dispersant concentration above the CMC results in a less pronounced decline in the oil-water interfacial surface tension because the oil-water interface is already occupied by a complete monolayer of surfactant molecules. The effectiveness of a chemical dispersant can be defined by the magnitude of the decrease in the oil-water interfacial surface tension to the CMC (i.e., effectiveness is greater when the magnitude of the decline in the interfacial surface tension to the CMC is greater). The efficiency of a chemical dispersant is defined as the concentration of dispersant required to reach the CMC (i.e., efficiency is greater when the CMC is attained at a lower concentration of dispersant). A preferred dispersant will have high degrees of both effectiveness and efficiency. However, the two concepts are not the same and do not necessarily follow similar trends (i.e.,

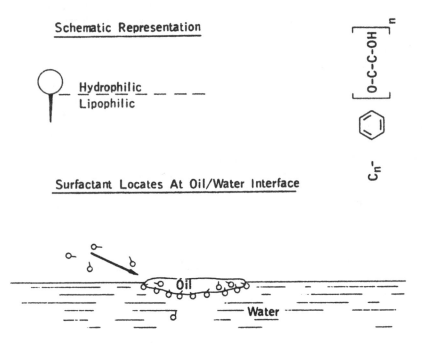

Figure 1. Orientation of surfactant molecule at the oil-water interface. *Source*: Canevari, 1978. (Copyright American Society for Testing and Materials, 1978. Reprinted with permission.)

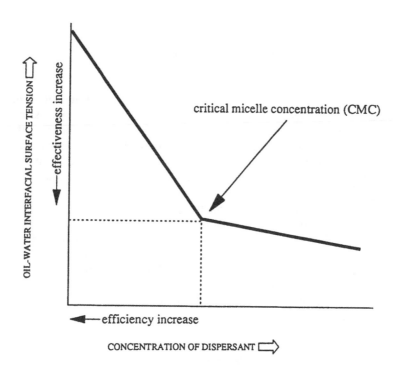

Figure 2. The relationship between oil-water interfacial surface tension, the concentration of a dispersant, and dispersant efficiency.

an effective dispersant may not be efficient, and vice versa).

The lowering of oil-water interfacial surface tension will promote dispersion of oil droplets into the underlying water with minimal mixing energy. As such, the total interfacial area for oil-water surfaces increases. Specifically, the relationship between mixing energy in a system, oil-water interfacial surface tension, and oil-water interfacial surface area is described by equation (1):

$$W_K = \gamma_{o/w} A_{o/w} \tag{1}$$

where

W_K = mixing energy,
$\gamma_{o/w}$ = oil-water interfacial surface tension, and
$A_{o/w}$ = oil-water interfacial surface area.

Thus, a lowering of the interfacial surface tension will promote the formation of a greater interfacial surface area (i.e., formation of more dispersed oil droplets) for a constant amount of mixing energy applied to the system. The presence of surfactant molecules at the oil-water interface also will tend to prevent reaggregation of oil droplets if they should encounter each other because of the common presence of hydrophilic groups of the surfactant molecules at the oil-water interfaces of droplets. The oil droplets will remain dispersed in a water column if they are small enough to allow for natural water currents or Brownian motion to prevent rising to reform surface slicks. Recent formulations for chemical dispersants actually can reduce interfacial surface tensions to such a level that only low amounts of mixing energy are required to promote interfacial film breakup and oil droplet formation. For example, Canevari et al. (1989) measured declines in oil-water interfacial tensions from 18 to less than 0.1 dynes/cm within 40 seconds of the addition of dispersants to Murban crude oil on seawater. The interfacial tension then gradually increased to 8-10 dynes/cm over approximately 60 minutes. Once an oil has been dispersed, even low levels of mixing energy should be sufficient to distribute and dilute the oil droplets in a water column.

As noted above, the true measure of the effectiveness of a chemical dispersant agent can be quantified by the decline in the oil-water interfacial surface tension to the CMC value. While the CMC value can be obtained in laboratory measurements such as the Drop-Weight test (see description in section on laboratory test methods), the latter method has inherent limitations that include (1) actual amounts of dispersed oil into a water column are difficult to estimate, (2) effects of relevant environmental variables (e.g., energy levels) on dispersion cannot be estimated, (3) accurate and reproducible measurements with the method require a relatively high degree of operator training and sophistication, (4) conditions in the method are far removed from real-world oil spill situations, and (5) deviations from application methods recommended by manufacturers of dispersant formulations are required to perform the test. In light of the preceding difficulties, an operational definition for dispersant effectiveness (or performance) is frequently based on analytical measurement of quantities of dispersed oil in a water column beneath a slick. Testing procedures in laboratory as well as field studies frequently utilize this operational approach to define dispersant effectiveness. It must be recognized, however, that many factors have pronounced influences on the quantities of oil that will be available for measurement in a water column beneath a slick. Much of the information in the following section addresses these factors and their influence on such operational measurements for dispersion effectiveness or, more accurately, performance.

Commercial formulations for chemical dispersants are comprised of two or more surfactant molecules that have various solubilities in both water and oils. One parameter that has been used to characterize these differential solubilities is the hydrophile-lipophile balance (HLB). The HLB is a coding scale that ranges from 0 to 20 for nonionic surfactants and takes into account the chemical structure of the surfactant molecule. A specific value for the HLB will characterize the tendency of the surfactant molecule to dissolve preferentially in either an oil phase (low HLB) or an aqueous phase (high HLB). The dominant group (i.e., lipophilic or hydrophilic) of a surfactant molecule will tend to be oriented in the external phase of an oil-water mixture. Therefore, a predominantly lipophilic surfactant (i.e., with a low HLB of 3 to 6) will favor water-in-oil emulsions (i.e., mousse). Natural surfactants, which tend to promote formation of mousse, are generally lipophilic in character. In contrast, a predominantly hydrophilic surfactant (i.e., with a high HLB of 8 to 18) will favor oil-in-water emulsions (i.e., dispersed oil droplets in a water body). The blend of surfactant molecules in commercial dispersant formulations tend to be of the latter type. These current formulations usually consist of a mixture of surfactant molecules with an overall HLB in the range of 9 to 11. Experiments have shown that HLB values in the latter range provide for more stable dispersions of oil droplets in water (e.g., Brochu et al., 1986/87).

The preceding information regarding the general action of chemical dispersant agents and its relationship to the lipophilic-hydrophilic character of surfactant molecules is simplified. In fact, the actual chemical dispersion process for oil in aqueous environments involves a complex set of interactions that depend on multiple processes and phenomena. Specifically, five requirements must be achieved for a chemical agent to successfully produce dispersion of oil into an underlying water column.

(1) The dispersant must be deposited onto the oil (i.e., onto the oil's surface) at the desired dosage.
(2) The dispersant (i.e., the surfactant molecules) must penetrate and mix with the oil.
(3) Surfactant molecules must orient at the oil-water interface with hydrophilic groups in the water phase and lipophilic (hydrophobic) groups in the oil phase.
(4) The oil-water interfacial surface tension must decrease due to the presence of the surfactant molecules at the oil-water interface, thereby weakening the cohesive strength of the oil film.
(5) Mixing energy must be applied at the oil-water interface, such as by wind and/or wave action at sea.

Critical factors affecting chemical dispersion of oil into water will include not only successful application of a dispersant onto an oil slick but also processes responsible for mixing of the dispersant into the oil, alignment of the surfactant molecules at the oil-water interface, dynamic reactions to which the surfactant molecules are subject at the oil-water interface, and subsequent partitioning and loss of surfactants from the oil into the water. The complex compositions of oils (and chemical dispersant formulations to a more limited extent) also will affect dispersant performance. Furthermore, oils immediately begin to experience rapid changes in their chemical composition and rheological and physical properties due to natural weathering processes following their release onto a water surface (e.g., Payne et al., 1983, 1984, 1991b, 1991c; Payne and McNabb, 1984; McAuliffe, 1989; Daling et al., 1990a). This weathering can lead to substantial changes in the responses to chemical dispersant agents. In summary, the concept of chemical dispersion of oil is relatively straightforward. In practice, the process and its associated measurement involves interplay between a series of complex interactions in both controlled laboratory and uncontrolled real-world (i.e., field) situations. Discussions of the effects of a number of important variables (i.e., chemical, environmental, and analytical) on dispersion of oil are presented in the section on factors affecting chemical dispersion, and the roles of these variables as they relate to operational measurements of dispersant effectiveness or performance (i.e., quantities of oil in water beneath a slick) in various laboratory tests are addressed in the section on laboratory testing.

CHEMICAL FORMULATION OF DISPERSANT AGENTS

As noted above, commercial formulations of chemical dispersant agents are normally comprised of mixtures of three types of chemicals: surface active agents (i.e., surfactants), solvents, and additives. As addressed in Farn (1983), initial dispersant formulations (i.e., in the 1950's) were based on highly aromatic solvents and non-biodegradable emulsifiers. While effective as dispersing agents, these formulations proved to be very toxic to a variety of marine organisms. In an effort to reduce toxicity, second-generation dispersant formulations began to appear in the late 1960's that were based on either hydrocarbon solvents with lower aromatic contents or water and a biodegradable, low toxicity emulsifier of the natural fatty acid polyglycol ester type. During the latter half of the 1970's, production also was undertaken for third-generation dispersant formulations. The latter were frequently produced in concentrated form (i.e., their application required dilution with ambient seawater) but maintained a relatively low toxicity for marine life. Non-aromatic hydrocarbons (or water-miscible solvents such as ethylene glycol or glycol ethers) and less toxic surfactants are currently used in formulations for recently developed dispersant formulations.

Exact compositions for commercial dispersant formulations are proprietary. However, chemical characteristics of the formulations are broadly known (e.g., Wells et al., 1985; Brochu et al., 1986/87; Fingas et al., 1990) and indicate that only a limited number of surfactant agents are currently used. Surfactants used in formulations can be grouped by charge type into four classes:

- Nonionic. These are the most commonly used surfactants in current formulations. Examples include sorbitan esters of oleic or lauric acids (e.g., sorbitan monooleate;

HLB=4.3; distributed as Span 80), ethoxylated sorbitan esters of oleic or lauric acid (e.g., ethoxylated sorbitan monooleate; HLB=15; distributed as Tween 80), polyethylene glycol esters of unsaturated fatty acids like oleic acid, ethoxylated or propoxylated fatty alcohols, and ethoxylated octylphenol.

- Anionic. Examples include sulfosuccinate esters (e.g., sodium dioctyl sulfosuccinate) as well as oxyalkylated C_{12} to C_{15} alcohols and their sulfonates (e.g., sodium ditridecanoyl sulfosuccinate).

- Cationic. An example is the quaternary ammonium salt $R(CH_3)_3N^+Cl^-$, where R is an organic moiety or part. Such compounds are not commonly used in current dispersant formulations because of their toxicity to a variety of organisms.

- Zwitterionic or amphoteric. Molecules of this type contain both positively and negatively charged groups that can produce a net neutral charge. An example would be a molecule containing both a quaternary ammonium group and a sulfonic acid group. Such compounds are not commonly used in current dispersant formulations.

Current dispersant formulations normally contain three surfactant molecules: two nonionic and one anionic (i.e., neither cationic nor zwitterionic/amphoteric types are in current formulations).

An example of how surfactant molecules will orient at an oil-water interface is presented in Figure 3. Compound A is sorbitan monooleate (HLB=4.3; predominantly lipophilic). Compound B is similar to A but has been ethoxylated with molecules of ethylene oxide to make it more hydrophilic (HLB=15). The dispersant formulation shown in the figure contains more of compound B than A. Such a balance will promote formation of oil-in-water emulsions because the dominant group of the surfactant mixture will tend to orient into the external phase (i.e., water surrounding a dispersed oil droplet). This is shown in the figure where the hydrophilic groups of compound B penetrate farther into the water phase. The use of two surfactants with differing HLB values but an overall average in the range of 9-11 allows for closer physical interactions (i.e., packing) of the surfactant molecules at the oil-water interface compared to a single surfactant with a HLB value in this range. This results in a stronger interfacial surfactant film and resistance to coalescence of dispersed oil drops. As noted above, dispersant formulations with an overall HLB in the range of 9 to 11 will generally yield the best dispersion of oil droplets in a water body. At the same time, the hydrophilic components of the surfactant molecules (especially B) will tend to promote the selective dissolution (i.e., partitioning) and loss of the surfactant molecules from the oil-water interface into the water over time. This eventual loss of surfactant to the water column should not affect the long-term stability of the oil dispersion, however, because of the dilution of the droplets to non-problematic concentrations in the water column.

The solvent portion of commercial dispersant formulations is intended primarily to not only dissolve solid surfactant molecules but also reduce the viscosity of the formulation to facilitate more uniform application and penetration of the dispersant onto and into an oil. Three

11

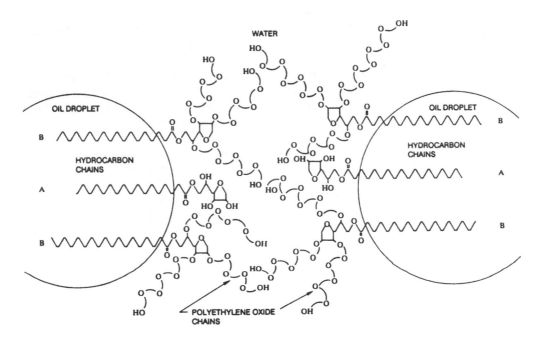

Figure 3.   Compound-specific orientation of two surfactants at the oil-water interface. *Source*: NRC, 1989. (Copyright National Academy of Sciences, 1989. Reprinted with permission.)

main classes of solvents can be used in dispersant formulations: (1) water, (2) water-miscible hydroxy compounds, and (3) hydrocarbons. Aqueous solvents (i.e., either water or appropriate hydroxy compounds) permit surfactants to be mixed directly into a water stream, which can facilitate applications of dispersants to an oil slick in field situations. Surfactant molecules (especially those with lower HLB values) will likely have greater solubilities in hydroxy-compound solvents than water. Examples of hydroxy solvents include ethylene glycol monobutyl ether, diethylene glycol monomethyl ether, and diethylene glycol monobutyl ether. Hydrocarbon solvents promote mixing and penetration of surfactants into oils that have higher viscosities from either their geochemical origin, the action of natural weathering processes on the oil, or low temperatures. Examples of hydrocarbon solvents include kerosenes with low aromatic content as well as branched saturated hydrocarbons that have higher boiling points but are less toxic than aromatics.

SECTION 3

FACTORS AFFECTING CHEMICAL DISPERSION OF OIL AND ITS MEASUREMENT

A variety of environmentally-relevant factors have major influences on the ability of chemical agents to disperse oil into water in both laboratory tests as well as actual field situations. A number of these factors are discussed in the following sections.

PROPERTIES AND CHEMISTRY OF OIL

Both crude and refined petroleum products are complex mixtures of hydrocarbon compounds. In a generalized characterization, crude and refined oils can be considered to contain compounds in five broad categories: lower molecular weight (1) aliphatics and (2) aromatics, and higher molecular weight (3) asphaltenes, (4) resins, and (5) waxes. Asphaltenes, resins, and waxes are defined in Bobra (1990). Asphaltenes are compounds that are soluble in aromatic solvents but insoluble in alkane solvents. They generally are considered to consist of condensed aromatic nuclei that contain alkyl and alicyclic systems with heteroatoms (e.g., nitrogen, oxygen, sulfur, metals, and salts) in various structural locations. Asphaltenes will have carbon numbers of 30 or greater and molecular weights of 500 to 10,000. Resins are complex higher molecular weight compounds containing oxygen, nitrogen, and sulfur atoms. They are polar and have strong adsorption tendencies. They normally will remain in solution following precipitation of asphaltenes (e.g., in alkane solvents) and will adsorb onto surface-active material (e.g., Fuller's earth). Molecular weights for resins are generally in the range of 800 to 1500. Waxes are hydrocarbon materials in asphaltene/resin-free oil that are insoluble in not only methylene chloride at 32°C but also methyl ethyl ketone. Petroleum waxes are also defined as higher molecular weight paraffinic substances (e.g., aliphatics) that crystallize out when an oil is cooled below its pour point. Petroleum wax is generally divided into two groups: paraffin wax and microcrystalline wax. Paraffin waxes are normal alkanes (n-alkanes) with carbon numbers of 20 to 40 and melting points of 32 to 71°C. Microcrystalline waxes mainly consist of iso-alkanes with carbon numbers of 35 to 75 and melting points of 54 to 93°C.

Interactions between aliphatics, aromatics, asphaltenes, resins, and waxes in complex oil mixtures allow for all of the compounds to be maintained in a liquid-oil state (Buist et al., 1989; Bobra, 1991). That is, the lower molecular weight components (i.e., the aliphatics and aromatics) act as solvents for the less soluble, higher molecular weight components (i.e., the asphaltenes, resins, and waxes). Such complex crude oil mixtures remain as relatively stable liquid phases as long as the solvency interactions occurring in the bulk oil are maintained and thermodynamic conditions remain constant. If this equilibrium state is changed, the solvency strength of the oil may become insufficient to keep the higher molecular weight components in solution and lead to their precipitation as solid particles. Accompanying changes in the physical state and chemical properties of the oil can affect the way chemical dispersants interact with the oil that has undergone such changes.

14

Values for physical and chemical properties for various oils are summarized in Tables 1 and 2 (from Clark and Brown, 1977). Triple dashes and single dashes in Tables 1 and 2, respectively, indicate no data is reported. Table 3 also summarizes information for chemical and physical properties of a variety of oils as well as values for dispersant performance with the premixed oil-dispersant version of Environment Canada's Swirling Flask test (see description in section for laboratory testing methods) and four chemical dispersants (Corexit 9527, Corexit CRX-8, Enersperse 700, and Dasic Slickgone). Values in Table 3 are from Bobra and Callaghan (1990) and Fingas et al. (1990). The three tables show that differences exist between oils for a number of physical, chemical, and rheological properties such as viscosity, pour point, specific gravity or density, oil-water interfacial surface tension, and chemical composition (e.g., general hydrocarbon classes as well as certain molecular elements). Furthermore, data in Table 3 illustrate that dispersant performance will vary as a function of not only oil type but also dispersant type. It is also important to note that oils have a certain inherent capacity for natural dispersion (i.e., dispersion that will occur independent of that caused by addition of a chemical dispersant if the oil is exposed to sufficient mixing energy). Fingas et al. (1989b) examined natural dispersibilities for fifteen types of oil. Tests were performed in the laboratory with two standard procedures used for evaluating dispersion performance: (1) the Labofina rotating flask test and (2) the Mackay or MNS test (see descriptions in section for laboratory testing methods). Both tests impart substantial amounts of turbulent energy to test solutions. Results are summarized in Figure 4. Data in the figure show that different oils are characterized by varying degrees of natural dispersibility that are a function of the testing procedure used to evaluate dispersion.

From the perspective of chemical dispersibility, the ability of particular petroleum products (i.e., either crude or refined oils) to be dispersed by commercial dispersants will vary as a function of physical and chemical properties of an oil. For example, oils characterized by higher viscosities will usually exhibit lower capacities for chemical dispersion. Specifically, increasing viscosity appears to reduce dispersion of oil droplets in two ways: (1) migration of the dispersant to the oil-water interface is retarded (i.e., the dispersant is unable to penetrate and homogeneously mix into a viscous oil) and (2) the energy required to shear off oil droplets from a slick is increased. Mackay and Wells (1983) note that there may be certain ranges of absolute viscosity (e.g., around 100 cpoise or cP) where an increase in viscosity may actually improve retention of dispersants by an oil and result in enhanced dispersant performance. However, it is generally accepted that chemical dispersants will perform better for oils with viscosities less than 2000 cStokes (cSt; note: absolute viscosity in cP = kinematic viscosity in cSt X density at a particular temperature) and essentially no dispersion will occur at viscosities greater than 10,000 cSt (Cormack et al., 1986/87) for oil on water. Dispersion data for the premixed version of Environment Canada's Swirling Flask test show agreement with these trends for viscosity (see Table 3). Oil-water interfacial surface tension provides the principal resistive force to formation of small droplets. Typically, the oil-water interfacial surface tension ranges from 20-30 dynes/cm for fresh oils. However, oil spill dispersants may reduce this value to 0.01 dyne/cm or less (Nes, 1984), which facilitates the natural process of droplet dispersion. Generally, sufficient mixing energy is present from waves in real-world situations (i.e., at sea) to form small,

Table 1. Analysis of crude and various boiling point cuts for Prudhoe Bay oil. *Source:* Clark and Brown, 1977. (Copyright Academic Press, Inc., 1977. Reprinted with permission.)

Characteristic or component	Unit	Crude oil	Natural gas	Naphtha		Middle Distillate	Wide-cut gas oils	Residuum
				Gasoline	Kerosine			
Boiling point range	°C	---	<20	20-190	190-205	205-343	343-565	565+
Specific gravity	(15°C)	0.8883	---	0.7531	0.818	0.8581	0.9279	1.0231
API gravity	°API	27.8	---	56.9	41.5	33.4	21.0	6.8
Pour point	°C	-10	---	---	<-60	-23	35	52
Viscosity, Saybolt (38°C)[a]	sec	73.5	---	---	---	36.1	85-200	>200
Kinematic (38°C)	cSt	14.0	---	---	---	3.05	>30	---
Yield: Crude oil	vol%	100	3.08	18.0	2.1	24.6	35.0	17.6
Paraffins	vol%	27.3	100	47.3	41.9	8.9	9.3	
Naphthenes	vol%	36.8	0	36.8	38.1	14.4	22.8	
Aromatics	vol%	25.3	0	15.9	20.0	76.7[b]	67.9[b]	
Others[c]	vol%	10.6	0	0	0			
Composition:								
Sulfur	wt%	0.94	---	0.011	0.04	0.34	1.05	2.30
Mercaptan sulfur	ppm	20	---	5	---	---	---	---
Nitrogen	wt%	0.23	---	0.02	0.02	0.04	0.16	0.68
Oxygen	wt%	0.01	---	---	---	---	---	---
Vanadium	ppm	18	0	0	0	0	<1	93
Nickel	ppm	10	0	0	0	0	<1	46
Iron	ppm	4	0	0	0	0	<1	25

a Saybolt viscosity: the time in seconds for 60 ml of a sample to flow through a calibrated Universal orifice under specified conditions, according to ASTM method D-88 [10:part 23].

b Includes naptheno-aromatic compounds and nonhydrocarbons.

c Polar compounds, non-volatile aromatic hydrocarbons, and column holdup in fractions boiling about 205°C.

Table 2. Physical characteristics and chemical properties of several crude oils. *Source:* Clark and Brown, 1977. (Copyright Academic Press, Inc., 1977. Reprinted with permission.)

Characteristic or component	Crude oil		
	Prudhoe Bay[a]	South Louisiana[b]	Kuwait[b]
API gravity (20°C) (°API)	27.8	34.5	31.4
Sulfur (wt%)	0.94	0.25	2.44
Nitrogen (wt%)	0.23	0.69	0.14
Nickel (ppm)	10	2.2	7.7
Vanadium (ppm)	20	1.9	28
Naphtha fraction[c] (wt%)	23.2	18.6	22.7
Paraffins	12.5	8.8	16.2
Naphthenes	7.4	7.7	4.1
Aromatics	3.2	2.1	2.4
Benzenes	0.3[d]	0.2	0.1
Toluene	0.6	0.4	0.4
C_8 aromatics	0.5	0.7	0.8
C_9 aromatics	0.06	0.5	0.6
C_{10} aromatics	-	0.2	0.3
C_{11} aromatics	-	0.1	0.1
High-boiling fraction[e] (wt%)	76.8	81.4	77.3
Saturates	14.4[f]	56.3	34.0
n-paraffins	5.8[g]	5.2	4.7
C_{11}	0.12	0.06	0.12
C_{12}	0.25	0.24	0.28
C_{13}	0.42	0.41	0.38
C_{14}	0.50	0.56	0.44
C_{15}	0.44	0.54	0.43
C_{16}	0.50	0.58	0.45
C_{17}	0.51	0.59	0.41
C_{18}	0.47	0.40	0.35
C_{19}	0.43	0.38	0.25
C_{20}	0.37	0.28	0.33
C_{21}	0.32	0.20	0.25
C_{22}	0.24	0.20	0.20
C_{23}	0.21	0.15	0.17
C_{24}	0.20	0.16	0.15
C_{25}	0.17	0.13	0.12
C_{26}	0.15	0.12	0.10
C_{27}	0.10	0.09	0.09
C_{28}	0.09	0.06	0.06
C_{29}	0.08	0.05	0.06
C_{30}	0.08	0.05	0.05
C_{31}	0.08	0.04	0.07
C_{32} plus	0.07	0.04	0.06
iso-paraffins	-	14.0	13.2
1-ring cycloparaffins	9.9	12.4	6.2
2-ring cycloparaffins	7.7	9.4	4.5
3-ring cycloparaffins	5.5	6.8	3.3
4-ring cycloparaffins	5.4	4.8	1.8
5-ring cycloparaffins	-	3.2	0.4
6-ring cycloparaffins	-	1.1	-
Aromatics (wt%)	25.0	16.5	21.9
Benzenes	7.0	3.9	4.8
Indans and tetralins	-	2.4	2.2
Dinaphtheno benzenes	-	2.9	2.0
Naphthalenes	9.9	1.3	0.7
Acenaphthenes	-	1.4	0.9
Phenanthrenes	3.1	0.9	0.3
Acenaphthalenes	-	2.8	1.5
Pyrenes	1.5	-	-
Chrysenes	-	-	0.2
Benzothiophenes	1.7	0.5	5.4
Dibenzothiophenes	1.3	0.4	3.3
Indanothiophenes	-	-	0.6
Polar materials[h] (wt%)	2.9	8.4	17.9
Insolubles[i] (wt%)	1.2	0.2	3.5

These analyses represent values for one typical crude oil from each of the geographical regions; variations in composition can be expected for oils produced from different formations or fields within each region.

a Adapted from Thompson et al. [23] and Coleman et al. [24].

b From Pancirov [25].

c Fraction boiling from 20° to 205°C.

d Reported for fraction boiling from 20° to 150°C.

e Fraction boiling above 205°C.

f Reported for fraction boiling above 220°C.

g Prudhoe Bay crude oil weathered two weeks to duplicate fractional distillation equivalent to approximately 205°C n-percents from gas chromatography over the range C_{11} to C_{32} plus for the Prudhoe Bay crude oil sample only. Unpublished data (R.C. Clark, Jr.)

h Polar material: Clay-gel separation according to ASTM method D-2007 [10; part 24] using pentane on unweathered sample.

i Insolubles: Pentane-insoluble materials according to ASTM method D-893 [10; part 23].

Table 3. Oil types and physical/chemical/dispersion properties

Oil Type	viscosities at 15C dynamic (cP)	kinematic (cSt)	density at 15C (g/mL)	hydrocarbon content (weight % of total) saturates	aromatics	polars	asphaltenes	waxes	oil-SW interfacial surface tension (dyn/cm)	premix Swirling Flask - dispersion performance (%) natural dispersibility	Corexit 9527	Corexit CRX-8	Enersperse 700	Dasic Slickgone
Adgo	61.6	64.6	0.9530	79.8	18.8	0.9	0.59	0.88	6.9	10	61	61	76	11
Amauligak	14	15.7	0.8896	89.5	9.3	0.4	0.31	0.87	20.9	8	50	61	65	23
Arabian Crude			0.8620				2.61	1.76		3	31	15	16	24
ASMB	9.2	11	0.8390	84.2	12.8	1.2	1.55	1.74	8.4	8	42	57	68	18
Atkinson	65.1	71.5	0.9110	82.7	13.2	1.5	2.39	0.72	17.9	8	59	67	79	33
Avalon (J-34)	11.4	13.5	0.8440	83.2	12.5	1.8	2.48	3.22	20.5	4	18	7.6	15	8
Bent Horn crude	24	29.3	0.8181	94.3	4.8	0.3	0.40	2.11	38.5	6	12	15	10	14
Bunker C	48000	48830	0.9830				6.73	1.23		0	2.3	3.8	0.9	2.1
California API=11	34000	34406	0.9882	13.7	29.8	31.4	18.63	2.37		0	0.5	2.3	0.4	0.2
California API=15	6400	6551	0.9770	13.7	36.4	24.1	20.13	1.60		0	1.3	0.4	0.9	0.8
Cohasset A-52	1.6	2.06	0.7900				0.35	0.90	16.5	6	88			
Cold Lake Bitumen	235000	234953	1.0002	16.6	39.2	24.9	11.87	1.35		0	1.9	1.1	0.9	1
Endicott	84	91.8	0.9149	87.1	10.9	1.3	3.16	0.54	25.8	3	17	20	10	8.1
Federated	4.5	5.4	0.8258	82.1	13.5	2.0	0.90	1.96	22.2	3	41	50	41	23
Hibernia	44.2	91.6	0.8849	91.5	2.7	0.3	3.62	1.10	13.5	4	13	14	7.3	8.6
Issungnak	3.1	3.652	0.8490				0.53	1.20	16.8	8	70	58	51	31
Lago Medio	41.1	47.1	0.8720				4.53	1.43	12.4	4	9.5	13	11	4.1
Lube Oil (used crankcase)	175.2	198	0.8848	86.3	12.9	0.8	0.00		21.0	17	42	39	47	29
Norman Wells	5.05	6.06	0.8320	86.3	11.1	1.6	1.15	1.25	16.4	5	51	60	73	19
Panuke F-99	1.1	1.47	0.7757	78.3	17.6	2.5	0.29	0.83		13	95	100	93	44
Prudhoe Bay	23	25.7	0.8936	65.1	26.3	8.4	2.04	0.65	27.4		19	23	48	14
South Lousiana			0.8390				0.20	1.06			53	55	31	27
Synthetic crude	4.6	5.3	0.8614	81.8	17.0	0.9	0.30	1.42	29.0	10	77	49	69	23
Terra Nova	22	25.7	0.8560				0.59	0.89	28.8	5	29	22	21	19
Transmountain	10.5	12.3	0.8550	81.0	13.6	1.9	3.50	1.39	19.3	3	14	13	17	11

information adapted from Bobra and Callaghan (1990) and Fingas et al. (1990)

18

oil type:

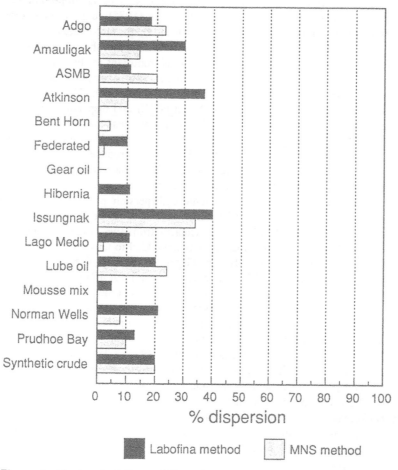

Figure 4. Mechanical dispersibility without chemical dispersants for
15 types of oil. Gear oil has no mechanical dispersibility
in either method. Data adapted from Fingas et al. (1989b).

dispersed oil droplets. Nevertheless, sufficient mixing energy must be provided to deform the oil, deform the water, and create new surface area for the oil. For low viscosity oils most of the mixing energy is consumed creating new surface area in the oil. For higher viscosity oils a relatively greater portion of the mixing energy will be consumed in deformation of the oil, which means that less energy will be available to promote formation of new surface area that results in dispersed oil droplets. Higher oil viscosities will also inhibit penetration and mixing of a chemical dispersant into an oil, which leads to lower dispersion of the oil. To summarize, new surface area of oil will be higher (i.e., resulting in more and smaller dispersed oil droplets) for low viscosity oils compared to more viscous oils.

The specific chemical composition of an oil can also be an important factor in the capacity for an oil to be chemically dispersed. Canevari (1985) suggests that some oils may contain surface active compounds that would allow the dispersant-treated oil to be dispersed regardless of its viscosity. However, Canevari also indicates that the overall composition and presence of indigenous surfactants in an oil may be as important as the viscosity for affecting dispersion in both increasing and decreasing manners. For example, dispersion tests were performed in the MNS test for two crude oils (La Rosa and Murban). The La Rosa and Murban crudes had kinematic viscosities of 73 and 6 cSt (at 60°C) and dispersion values of 30% and 78%, respectively. Initial appraisal of these data might suggest that differences in dispersion could be attributed to the viscosity differences between the oils. However, the La Rosa crude was mixed with a pure isoparaffin oil of low viscosity to produce a modified crude with a viscosity similar to that of the Murban crude. This modified crude had a dispersion of only 50% (versus 78% for the Murban crude). This illustrates that characteristics of an oil other than viscosity alone can be important for the chemical dispersion process. In further studies, Canevari (1987) investigated the role of natural surfactants in oils for influencing chemical dispersion. Natural surfactant fractions were isolated from five crude oils (i.e., Kuwait, La Rosa, North Slope, Murban, and South Louisiana) and added to tetradecane (nC_{14}) in amounts equivalent to those in the parent crude oils. The dispersion for nC_{14} (i.e., without the surfactant additives from the crude oils) was 46%. However, dispersion for nC_{14} was reduced to 13-20% following the additions of the natural surfactants from the five crude oils. Fingas et al. (1991b) report that chemical dispersion of oils increases with increasing saturate content and declines with increasing content of aromatic, polar, and asphaltene components. Wax content of oils also can be important in the formation and stabilization of water-in-oil emulsions (or mousse) for oil released at sea (e.g., Bridie et al., 1980; Bobra, 1990, 1991). Thus, high wax content could have a negative effect on chemical dispersion of emulsified oils. Further studies on relationships between dispersant performance and the chemical composition of oils are warranted because studies to date have not found identical trends. For example, a positive relationship between dispersant performance and asphaltene content (i.e., increasing dispersion with increasing asphaltene content) for oils with 5% or less asphaltenes by weight has been found at IKU Sintef Gruppen (P.S. Daling, personal communication).

In addition to inherent differences in chemical compositions of different parent crudes and refined products, an oil that is released onto a water's surface or into a water column will undergo a variety of rapid, dynamic changes in both its chemical composition and physical properties due to natural weathering processes (Jordan and Payne, 1980; Payne and McNabb, 1984; Payne et al., 1983, 1984, 1991c; Payne and Phillips, 1985; McAuliffe, 1989; Daling et al., 1990a). For example, many of the lower molecular weight components (i.e., aliphatics and aromatics) that are important for the solvency interactions for higher molecular weight components (i.e., waxes, resins, and asphaltenes) are selectively lost from an oil due to evaporation and dissolution during natural weathering processes. Furthermore, water is rapidly incorporated into many oils to form stable water-in-oil emulsions (i.e., mousse), which are characterized by lower oil-water interfacial surface tensions and substantially higher viscosities (e.g., greater than 2000 cP; Payne et al., 1983; Cormack et al., 1986/87; Daling et al., 1990a). Examples of the latter trends are shown in Figure 5 (from Payne et al., 1983, 1991b). Studies have shown that viscous mousse is more resistant to chemical dispersion than its less viscous parent oil (e.g., Buist and Ross, 1986/87; Daling, 1988; NRC, 1989; Daling et al., 1990a). Formation of water-in-oil emulsions generally occurs rapidly in field situations and is, therefore, important in considerations for chemical dispersion of oil in spill events. That is, the period of time or window-of-opportunity during which chemical dispersants can be effectively utilized for a spill event may be relatively short (e.g., only a matter of days) before water-in-oil emulsification can effectively preclude the successful dispersion of an oil by chemical agents.

A number of investigators have shown that the starting composition of a parent oil can have a major influence on its predisposition to form stable water-in-oil emulsions (mousse). For example, the presence of natural surfactants in the wax, resin, and asphaltene fractions of oils has been positively correlated with the tendency to form mousse. Bridie et al. (1980) was able to obtain a stable mousse by stirring the greater than 200°C (i.e., 200+) fraction of Kuwait crude in seawater. No stable mousse was formed in an identical experiment following removal of the asphaltene and wax fractions from the parent 200+ oil. Addition of the combined asphaltene and wax fractions from the 200+ oil to a lubricating base oil also caused the latter to form a stable mousse with stirring in seawater. The lubricating oil by itself had no tendency to form such a mousse. Additional laboratory studies have shown that mousse formation is favored in the presence of metalloporphyrins, which are naturally occurring components with surface-active properties in crude oils (NRC, 1985). Canevari (1982) noted that two mechanisms may be operative to stabilize water-in-oil emulsions: (1) actions of natural surfactants such as those addressed in Bridie et al. (1980) and NRC (1985) and (2) the presence of bi-wetted solid particles (i.e., partly water-wetted and partly oil-wetted) at oil-water interfaces that prevent emulsified water droplets in oil from coalescing with each other. Bobra (1990, 1991) considers the latter point in greater detail and proposes that submicron-size particles of higher molecular weight components in oil (i.e., asphaltenes, waxes, and resins) actually stabilize water-in-oil emulsions. Bobra maintains that the loss of the lower molecular weight aliphatics and aromatics (i.e., the solvents for maintaining asphaltenes, waxes, and resins in solution in the parent oil) during natural weathering processes will likely result in additional precipitation of the higher molecular weight components and an accompanying enhancement of water-in-oil emulsification. Bobra (1990) also notes that factors such as the alkane/aromatic ratio and the compositions of the alkane and aromatic fractions in an oil can affect the tendency for precipitation of asphaltenes (i.e., asphaltenes are soluble in aromatic solvents and insoluble in alkane solvents) and, therefore, the oil's emulsification behavior.

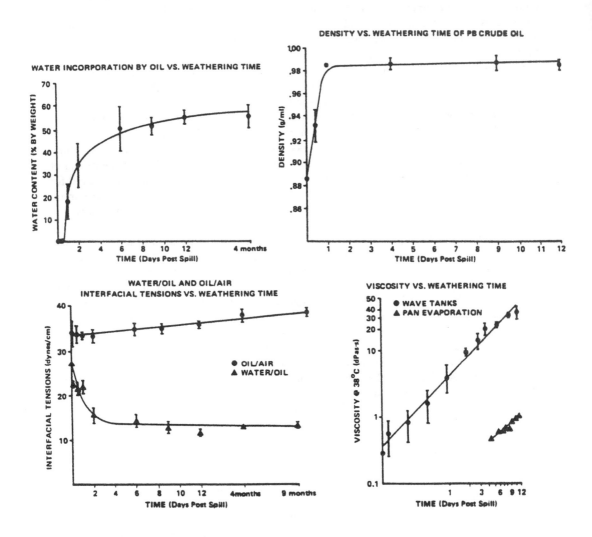

Figure 5. Rheological properties in Prudhoe Bay crude oil with natural weathering over time in outdoor, flow-through seawater wave tanks. Points are means ± one standard deviation from three wave tanks. *Source*: Payne et al., 1991b. (Copyright American Petroleum Institute, 1991. Reprinted with permission.)

22

Both water-in-oil emulsification (i.e., mousse formation) and oil-in-water dispersion (i.e., due to chemical dispersants) result from the actions of surface active agents in oil. However, the two processes yield essentially opposite and incompatible products. As noted above, the formation of one tends to inhibit formation of the other (i.e., the natural process of water-in-oil emulsification can yield a very viscous mousse that is resistant to chemical dispersion). Because mousse formation can rapidly occur in natural spill events, attempts have been made to formulate chemical agents (i.e., demulsifiers) that will counteract or break water-in-oil emulsions. Such efforts are addressed in the work of Bridie et al. (1980), Canevari (1982), and Buist and Ross (1986/87), although exact compositions of the demulsifier agents are not provided for proprietary reasons.

DISPERSANT COMPOSITION

Numerous studies have shown that the degree of dispersion for a particular oil in a particular laboratory test will vary with different chemical dispersants (e.g., Canevari, 1985, 1987; Desmarquest et al., 1985; Fingas et al., 1989a and b, 1990, 1991a and b; Daling et al., 1990b). Examples of this have been shown in Table 3. Laboratory studies conducted in conjunction with the preparation of this book also evaluated dispersion performance with several commercial agents (Clayton et al., 1992). The tests were performed with three dispersants: Corexit 9527 (C9527), Corexit CRX-8 (CRX-8), and Enersperse 700 (EN700). Controls without a dispersant agent also were included in the studies. Results were obtained for two oils (Prudhoe Bay crude and South Louisiana crude) in three testing procedures (10-minute and 2-hour samples in the Revised Standard EPA test; premixed, 1-drop, and 2-drop dispersant-to-oil versions of the Swirling Flask test; and the IFP-Dilution test; see Section 4 for descriptions of the procedures). Results of dispersion performance are illustrated in Figure 6. As shown, dispersion in the presence of all three chemical agents was greater than that in the no-dispersant controls. However, the performance values did differ among dispersant agents as functions of both the testing procedure and the type of oil used.

As noted above, differences in dispersant formulations can influence the degree of dispersion of oils. Commercially-available dispersants are comprised of different mixtures of surface active compounds (i.e., surfactants), solvents, and additives. While exact compositions of formulations are proprietary, certain characteristics of the formulations and how they relate to dispersion of oil are broadly known (e.g., Brochu et al., 1986/87; Wells et al., 1985). Surfactants are compounds containing both lipophilic and hydrophilic groups, which allows the molecules to reside at oil-water interfaces and lower oil-water interfacial surface tension. Each surfactant is characterized by an HLB (i.e., hydrophile-lipophile balance) value, with the range of values in useful dispersant formulations varying from 1 (most lipophilic) to 20 (most hydrophilic) for nonionic surfactants. Current dispersant formulations usually have an overall HLB of 9 to 11, which derives from a mixture of lipophilic and hydrophilic surfactants with HLBs of approximately 5 and 15, respectively. This has been shown to provide for more stable chemical dispersions of oil droplets in water (e.g., Brochu et al., 1986/87). It should be noted, however, that the overall HLB value of surfactants in a chemically-dispersed oil can change

23

(a) Revised Standard EPA test---10 minute sample

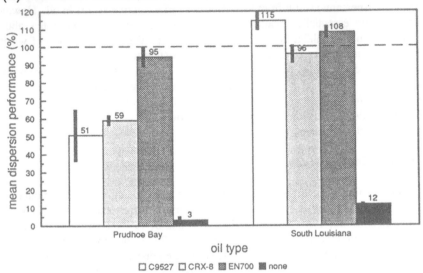

(b) Revised Standard EPA test---2 hour sample

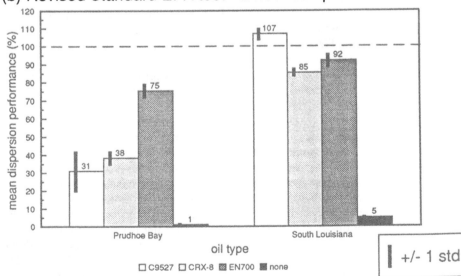

Figure 6. Dispersant performance for separate testing procedures with two oils, three dispersants, and a "no dispersant" control. Values are means and +/- 1 standard deviation (n=3 to 13).

(c) Swirling Flask test---premixed version

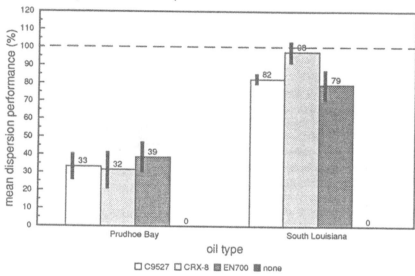

(d) IFP-Dilution test

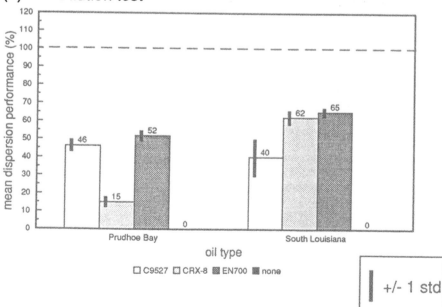

Figure 6. (continued)

(e) Swirling Flask test---1-drop version

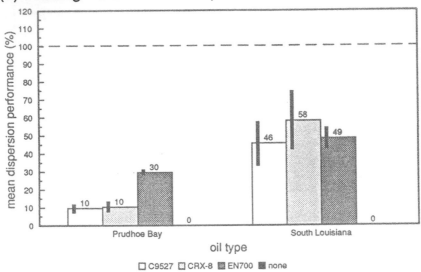

(f) Swirling Flask test---2-drop version

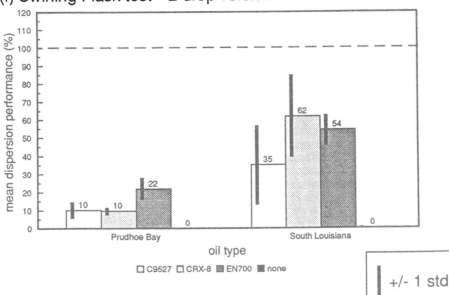

Figure 6. (continued)

(e.g., decrease) over time as the hydrophilic surfactant(s) selectively dissolve or partition from the oil-water interface into the water. Temporal changes in the quantity and type of surfactants present at the oil-water interface can affect the chemically-dispersed character and behavior of the oil. The solvent portions of commercial dispersant formulations can consist of water, water-miscible hydroxy compounds, or hydrocarbons. The purpose of the solvent is primarily to dissolve solid surfactant molecules and reduce the viscosity of the final formulation to facilitate more uniform application and penetration of the dispersant onto/into an oil. Additives may be present in dispersant formulations for purposes such as increasing the biodegradability of dispersed oil, improving the dissolution of the dispersant into an oil slick, and increasing the long-term stability of a dispersion.

In addition to the effects of different dispersants on resulting values of dispersion of oils, the composition of a dispersant formulation also may influence the composition of an oil during the dispersion process by chemical fractionation. For example, Fingas et al. (1989c) present data from both gravimetric and flame-ionization-detector gas-chromatographic (FID-GC) measurements that show increased evaporative losses of lower molecular weight components from chemically dispersed oil. This increased rate of weathering presumably reflects more rapid dissolution and evaporation losses of more water-soluble and volatile compounds through the increased surface area of the small dispersed oil droplets. The FID-GC information also suggests that the specific composition of the surfactants can affect the composition of the oil. For example, surfactant molecules may selectively draw oil compounds corresponding to their lipophilic chain lengths into the water (i.e., these oil compounds become relatively enriched in the water and depleted in the oil). Furthermore, oil compounds with longer carbon-chain lengths (i.e., greater than the lipophilic chain lengths of the surfactant molecules) may become selectively enriched in non-dispersed oil that remains at the water's surface.

In general, the following criteria must be satisfied for chemical dispersion of oil:

- the dispersant must be successfully and uniformly applied to a slick,

- the surfactant molecules in the dispersant must mix with the oil and move to the oil-water interface,

- the surfactant molecules must attain a concentration at the oil-water interface that causes a sufficiently large reduction in the oil-water interfacial surface tension,

- the surfactant molecules must remain at the oil-water interface for a sufficient period of time to allow for the formation (i.e., dispersion) of oil droplets, and

- the oil must be dispersed into droplets, which requires an input of mixing energy to the system.

Variations in the composition of dispersant formulations can have particular impact on the second through the fourth of these criteria, as evidenced in the studies of Brochu et al.

27

(1986/87).

Once oil droplets of sufficiently small sizes have been dispersed into a large body of water, subsequent dissolution losses of selected surfactant molecules (e.g., hydrophilic types) from the oil-water interface into the water will be inconsequential to the final dispersion result. This is because the oil droplets will remain in suspension (i.e., not rise back to the air-water interface due to their small sizes) and have sufficiently low spatial densities so that they will not interact (i.e., coagulate) with each other.

DISPERSANT APPLICATION

Major factors affecting dispersion of oil in laboratory and field studies include (1) the method of application of a dispersant to an oil, (2) the method of mixing of the dispersant into the oil, and (3) the subsequent dispersion of the dispersant/oil mixture into the water column. The dispersant application method can be one of the most critical elements determining whether a particular dispersant will produce dispersion of oil or not. In field situations, dispersants are normally applied to oil slicks on water from aircraft (airplanes or helicopters) or surface vessels (boats). From all of the preceding platforms, dispersants are released from apparatus designed to dispense the chemical formulation as relatively small droplets that will descend onto a slick in a manner providing broad spatial coverage. Although hand-held applicators (e.g., back-pack spray units) also might be used, the aerial extent of their coverage necessarily will be much more limited. The size of the applied dispersant droplets will be important to successful application. Droplets that are to large may penetrate through an oil slick without interaction. Droplets that are to small may not reach the slick because of air or wind transport between the application source and the slick. As discussed below, application technology and practice are critical and will vary depending on the ultimate purpose in laboratory or field efforts.

Two phenomena related to interactions between dispersant droplets and an oil slick can affect dispersion: (1) herding reactions and (2) roll-off effects (e.g., Canevari, 1984, 1991; Fingas, 1989). Herding will tend to occur in thinner slicks where the dispersant droplets interact directly with the water surface (e.g., dispersant droplets are applied directly onto the water or penetrate completely through a slick). Surfactant molecules that are introduced onto the water surface have a driving force (or spreading pressure) to orient themselves based on their hydrophilic-lipophilic components. As a result, dispersant molecules at the water surface may push or herd the oil aside and lead to a decline in oil dispersion because dispersant does not gain access to the oil. At the same time, intentional herding of oil might be useful in field situations for concentrating oil into thicker herded slicks, which would be more amenable to a second-pass application of dispersant from aircraft or boat platforms. Fingas et al. (1990) suggest that the tendency for herding may increase progressively as the HLB of surfactant molecules increases above 10. This can result in an increased disposition toward herding over time for chemically-treated oil because the hydrophilic surfactant(s) (i.e., higher HLB) of a dispersant formulation will preferentially dissolve into the water phase where they can orient at the air-water interface and promote herding. The rate of such a selective leaching of hydrophilic surfactants into a water phase relative to the actual occurrence of herding for a slick warrants further study (P.S.

Daling, personal communication).

Roll-off of dispersant from oil can occur with viscous oils, where dispersant droplets may have difficulty penetrating into the oil slick and simply roll-off of the oil into the water. The latter phenomenon will lead to decreased interaction between surfactant molecules and oil, which will result in reduced dispersion.

Three approaches have been used for dispersant applications in laboratory tests: (1) premixing of a dispersant with an oil before a test begins, (2) premixing of a dispersant with the water before oil is introduced to the system (Rewick et al., 1981, 1984), and (3) mixing of the dispersant with the oil at the oil-air interface as part of the testing procedure itself. Most tests in which a dispersant is premixed with an oil have indicated that adequate dispersion of oil can be obtained. Results from such tests are indicative of the dispersion that can be imparted by a particular dispersant to a given oil because artifacts associated with inefficient application or mixing of the dispersant onto and into the oil slick are eliminated. Consequently, premixing of a dispersant with an oil is attractive when the purpose of testing is to compare or rank dispersibilities imparted by different dispersants to an oil. It should be recognized, however, that premixing of a dispersant with an oil is not representative of conditions at real-world spill events. In those laboratory tests where dispersant is added to an oil slick as an integral part of the testing process itself (i.e., the dispersant is applied dropwise or sprayed onto the slick), energy for mixing of the dispersant into the oil is concentrated at the oil-air interface. The latter situation is more likely to simulate conditions that could be encountered at sea and ensures that dilution of a dispersant into an oil is more gradual. Additional factors related to dispersant application that can affect results of dispersion measurements include whether the dispersant is applied in a neat or diluted form and the droplet size of the applied dispersant formulation.

Mackay et al. (1984) discuss various aspects of the ratio of dispersant drop diameter to oil slick thickness, and indicate that dispersant drop diameters of 500 um (plus or minus a factor of 3) generally would exceed a normal slick thickness (200 um plus or minus a factor of 5). At droplet sizes greater than 500 um, the dispersant could penetrate through the slick and cause herding of the oil due to concentration of surfactant molecules at the air-water interface and an accompanying lowering of the water-dispersant-air interfacial surface tension. Thus, it would appear that improved dispersion performance could be achieved with dispersant droplets having smaller sizes. However, smaller droplets applied by aircraft in field situations are more subject to wind drift, which can lead to decreases in application effectiveness. The density and viscosity of dispersant droplets during application also are important because the dispersant can migrate rapidly through a thin, low viscosity oil slick. Dispersant droplets may not migrate into heavier, more viscous oils as quickly, and can be washed off the oil surface and lost to the surrounding water. Tendencies to either penetrate into or wash off of a slick will depend on the chemical compositions of both the dispersant formulation and the oil, as well as the degree of natural weathering that has occurred in the oil. The median droplet diameter for a dispersant application should be at least 400 um for achieving an adequate deposition density on a slick, regardless of the dispersant application method. Meeks (1980) found that the thickness of a slick and the soaking-time (i.e., the time that the dispersant could mix with the oil slick before the onset of agitation) could substantially affect dispersion.

Belore (1987) conducted studies on dispersion in a 125-L tank. Mixing energy was provided to the system with an oscillating hoop that tended to keep oil on the water's surface in the center of the container. An apparatus for dispersant delivery to the test tank provided droplets of dispersant with uniform distributions in the following three sizes: 300 um, 650 um, and 1100 um. Tests of oil dispersion with the system were conducted for three dispersants (Corexit 9527, Enersperse 700 or EN700, and Finasol OSR5) and four crude oils (Alberta Sweet Mixed Blend, Uviluk, Norman Wells, and Redwater) with the three sizes for dispersant droplets and varying thicknesses of oil slicks. Although not always conclusive, the results of a number of the tests (e.g., especially EN700/Norman Wells crude and Corexit 9527/Uviluk crude) gave evidence that oil dispersion increased as the droplet size of dispersants decreased.

Smedley (1981) studied the effect of dispersant droplet size on dispersion. These studies also were performed with an oscillating hoop test in a 30 cm glass tank equipped with a 25 cm stainless steel ring that oscillated at the water surface at a rate of 350 cycles/minute. The radial wave pattern tended to promote accumulation of the oil (Alberta crude) at the center of the water's surface in the test tank. The dispersant (Corexit 9527) was then sprayed through an air atomizing nozzle onto the oil at a dispersant-to-oil ratio of 1:20. Droplet size was determined by spraying a volume of dyed dispersant onto a moving paper target and then counting and measuring the droplet stains. The water was agitated for 30 minutes after dispersant application, followed by sampling and analysis by solvent extraction and infrared (IR) detection of dispersed oil. Variables in the study included mean size of dispersant droplets (i.e., drop mass mean diameters of 400-1300 um) as well as thickness of oil films. Thicker oil slicks were obtained by adding more oil to the test tank, which required an accompanying application of additional dispersant. Three thicknesses of oil slicks were tested: 400 um, 800 um, and 1600 um. Results suggest that larger dispersant droplets produced lower dispersion of the oil, although these trends were not large. If real, greater penetration of the larger droplets completely through the slick presumably resulted in reduced dispersant-oil interactions. With viscous or weathered oils, larger drop sizes of dispersants may be more likely to produce dispersion because of their increased tendency to be retained by the slick.

Smedley (1981) also discusses additional factors that can affect dispersant applications from aircraft. Such factors include atmospheric transport of dispersant droplets (including the aircraft vortex effect on the terminal settling velocities for the droplets), the actual velocities of the droplets, atmospheric conditions including wind and turbulence, and the roughness of the ocean's surface. Complicating factors arising during aerial applications of dispersants also are discussed in Lindblom and Cashion (1983), Becker and Lindblom (1983), Dennis and Steelman (1980), and Lindblom and Barker (1978). As noted in Merlin et al. (1989), spray application of dispersants (by aircraft or boat) in a downwind direction can produce reduced dispersion for an oil slick. This results from wind-aided transport and deposition of the smaller dispersant droplets in front of the main dispersant-application path, producing herding of the oil slick. Upon arrival of the main application dose for the dispersant, the herded oil will be less effectively dispersed because the bulk of the applied dispersant may not contact the oil. The preceding complications pertaining to effective application of dispersants onto an oil slick emphasize a major difficulty that frequently influences chemical dispersion of slicks. Specifically, dispersion of a slick may be largely ineffective because of inadequate application of dispersant onto the slick, regardless of the capacity of the dispersant for dispersing the oil (i.e., lowering the oil-water interfacial surface tension).

Even in light of the preceding complications, aerial applications of dispersants remain a necessary and essential component for any oil-spill contingency plan, particularly for remote areas. As such, studies to further the understanding of affects of dispersant delivery conditions (e.g., application of dispersant droplets in a sprayed form) as well as additional development of delivery systems and technology need to be performed.

MIXING ENERGY

For the chemical dispersion process, oil-water interfacial surface tension values are first lowered as a result of mixing of the dispersant with the oil and movement of surfactant molecules to the oil-water interface. The actual dispersion of the oil as droplets into a water column then requires an input of mixing energy to the system. Following their injection into the water, the dispersed oil droplets are subject to advective processes that result in horizontal and vertical transport and dilution. Dispersion in the vertical plane is countered by the buoyancy of the oil droplets, which depends on the density and size of the droplets. Rise velocities for oil droplets can be described by Stokes' Law (2):

$$v_r = [gd^2(\rho_w - \rho_o)]/18\eta \qquad\qquad (2)$$

where v_r = rise velocity,
   g = gravitational constant,
   d = diameter of the oil droplet,
   ρ_w = density of the aqueous fluid medium,
   ρ_o = density of the oil droplet, and
   η = viscosity of the aqueous fluid medium.

As can be seen, rise velocities will be a function of the square of the diameter of an oil droplet. The size of dispersed oil droplets generated during mixing will have an important bearing on estimates of dispersion that are based on experimental measurements of oil (droplets) in a water column. Likewise, density differences between the oil and water will affect the rise velocities of the oil droplets. If the influence of advective transport processes affecting the dispersed oil droplets exceed rise velocities or times for the droplets, then the dispersed oil will remain in suspension in the water column.

For dispersion to be effective, oil droplets must remain in suspension and not return to the water's surface to reform a slick. Consequently, some laboratory tests include an initial period of agitation to disperse oil droplets and then a static or stationary period in which those droplets that have a higher buoyancy (e.g., larger droplets that will not remain as a relatively stable dispersed suspension) are allowed to return to the surface of the test solution. Subsurface samples for analysis of dispersed oil (droplets) remaining in suspension are subsequently collected after a specified settling-time or stationary period.

A number of investigators have studied the relationship between mixing energy and the sizes of oil droplets generated in solutions. Delvigne (1987, 1989) and Delvigne and Sweeney (1988) demonstrated that droplet size distributions for naturally dispersed oil (i.e., in the absence of chemical dispersants) become smaller as both the magnitude and duration of mixing energy increases. Increasing oil viscosity (as influenced by the oil type, its weathering state, and temperature) will tend to produce larger dispersed oil droplets under equivalent mixing energy regimes. Jasper et al. (1978) demonstrated that sizes of oil droplets decreased for both non-chemically and chemically dispersed oil as energy input increased in stirred tank experiments. Lewis et al. (1985) and Byford et al. (1984) determined volume mean diameters (vmd) for dispersed oil droplets (i.e., vmd is the droplet diameter at which half the total volume of dispersed oil is contained in droplets with diameters less than the vmd) as a function of varying energy levels with several standard laboratory testing methods for evaluating dispersion. Tests were performed with two oils (Lago Medio +200°C residue and a medium fuel oil) and three unidentified dispersant formulations (one experimental and two commercially available). The results demonstrated that increasing turbulent energies applied to systems produced not only smaller vmd values for oil droplets but also greater dispersion of the oils in the test methods. An example of the relationship between dispersion and vmd is shown in Figure 7 (Lewis et al., 1985). A similar correlation between increasing turbulent energies and greater dispersion for chemically-treated medium fuel oil is reported in Lee et al. (1981). Consequently, data do show that dispersion of chemically treated oil can be positively correlated with the level of energy introduced to a system and negatively correlated with the resulting size of the dispersed oil droplets. Fingas et al. (1991b) report that truly stable oil-in-water dispersions occur when the distribution of oil droplets has a vmd of approximately 30 um, regardless of the laboratory method and source of mixing energy used to generate the dispersion. Daling et al. (1990a and b) report that the relationship between dispersion and oil droplet diameters (e.g., the diameters corresponding to cumulative volume distributions of 10%, 50%, and 90%) can be more complex. The stability of dispersions can be functions of variables such as the type of oil and dispersant tested, the weathered state of the oil, the volume ratio of dispersant to oil, the test method used to determine the degree of dispersion, the densities of oil droplets generated in the test, and the rates of coalescence of oil droplets in the test. Daling et al. conclude that stable dispersions may not always have vmd values in the range of 30 um.

As discussed in greater detail in the Section 4, a variety of methods have been devised to introduce agitation into procedures designed to measure dispersion performance in the laboratory. Specifically, agitation has been introduced to test systems by the following mechanisms:

- Shaking, swirling, inverting, or rotating of flask systems (e.g., Labofina, Warren Spring Laboratory, and Environment Canada tests).

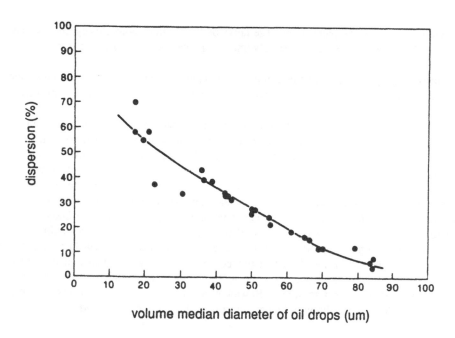

Figure 7. Relationship between dispersant performance (Labofina rotating flask test) and volume-mean-diameter (vmd) of dispersed oil drops for Lago Medio +200 crude oil at 20°C. *Source*: Lewis et al., 1985. (Copyright American Petroleum Institute, 1985. Reprinted with permission.)

- Activation of stirrers or oscillating paddles, plates, or rings (e.g., IFP-Dilution and oscillating hoop tests).

- Spraying and recycling of test solutions through circulating pump systems (e.g., Revised Standard EPA test).

- Generation of wave action by passage of air over a solution in the test apparatus (e.g., MNS test).

- Generation of wave patterns in large tanks or flumes that more closely approximate real-world situations at sea (e.g., Delft Hydraulics and cascading weir flume tests).

While providing agitation, all of these approaches have inherent properties and limitations that can influence dispersion performance as well as making comparisons among tests problematic.

For example, the presence of oil on the water's surface can lead to a dampening of wave activity in test systems such as those utilizing an air current over the water to generate mixing energy (i.e., the MNS test). Wave dampening has been discussed at considerable length by Mackay and Szeto (1982), Mackay et al. (1983), Mackay et al. (1984), and others. Studies reported in those papers note that the wave height on the water's surface is reduced (i.e., wave dampening) with No. 6 fuel oil and some dispersant oil combinations. Reduction in wave height means that less turbulent energy is being generated in the experimental water body, which leads to decreases in the amount of oil transferred (i.e., dispersed) into the water column. Such a situation can lead to poor reproducibility in test results. Rewick et al. (1981) reported that the amount of oil dispersed in the MNS test procedure is generally less than that observed in the Revised Standard EPA test (i.e., a tank test involving generation of mixing energy by recycling/spraying of the test solution). Rewick et al. state that the level of mixing energy may be lower in the MNS test, although higher control-test blanks (i.e., 37% and 31% dispersion for No. 2 and No. 6 fuel oils, respectively, in the absence of chemical dispersants) relative to equivalent values in the EPA test (6.5 and 0.9% dispersion, respectively) suggest that mixing energy may actually be higher in the MNS test. An alternative explanation is that the factor of two difference in the oil-to-seawater ratios (OWR) for the two tests (i.e., OWR values in the MNS and EPA tests were 1:600 and 1:1300 v:v, respectively) may have contributed to the higher control blanks in the MNS test.

Byford and Green (1984) compare dispersion results from the MNS and Labofina tests, which have different mixing regimes. The authors found good correlation between results with the two tests using a variety of commercial dispersants and weathered Kuwait crude oil at 10°C (Figure 8). It is important to note that comparable test results were produced with the two methods despite differences in dispersant-to-oil ratios, oil-to-seawater ratios, methods of agitation, and methods of sampling. The ranking order for dispersant performance for both tests was consistent and values of absolute dispersion performance were similar. It was believed, however, that this similarity could have been due to the particular oil used, whereas other oils

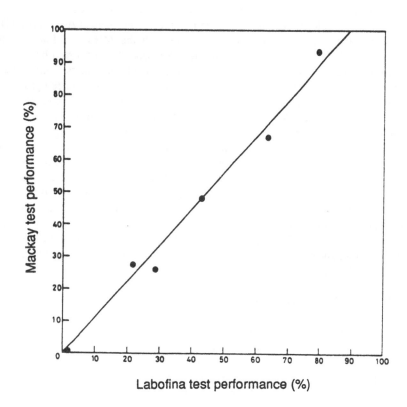

Figure 8. Correlation between dispersant performance for weathered Kuwait crude oil at 10°C with the MNS (Mackay) and Labofina tests. *Source*: Byford and Green, 1984. (Copyright American Society for Testing and Materials, 1984. Reprinted with permission.)

might yield different values for dispersion performance and performance rankings by dispersant. Also, those results from experiments in which wave dampening occurred in the MNS apparatus were not included in the comparison. Byford and Green also conducted tests with combinations of various dispersant formulations to evaluate possible synergistic effects between dispersant mixtures. Results indicated that the choice of a dispersant mixture yielding the highest dispersion value might or might not be the same with the two testing methods. For example, similar dispersion results were obtained for weathered Kuwait crude with the combinations of surfactants C+D (Figure 9a). However, the dispersant mixture giving the highest dispersion value with surfactants A+B was slightly different with the two procedures (Figure 9b). When weathered Lago Medio crude was tested, the mixtures of surfactants A+B and C+D yielding the highest values for dispersion were quite different with the MNS and Labofina test methods (Figures 9c and d). Wave dampening with the MNS procedure occurred before the end of the tests with weathered Kuwait crude and all combinations of surfactants C+D. When weathered Kuwait crude was treated with surfactants A+B, wave dampening occurred progressively earlier as the proportion of surfactant A was increased relative to B. Another problem encountered by Byford and Green with the MNS test procedure was that values for dispersion as high as 123% were obtained on occasion. These occurrences suggested that nonhomogeneous distributions of dispersed oil (i.e., due to large drops and mechanical dispersion independent of that resulting from the chemical dispersion process) were generated and maintained to some degree by wave motion. Larger, unstable droplets of oil could have been collected because samples were withdrawn from the test apparatus while wave action was still being applied to the test solution. This aspect of the MNS procedure as applied by Byford and Green could result in artificially high estimates for dispersion and make comparison between results with different methods problematic. Byford and Green concluded that the Labofina test method did not suffer from a similar limitation, and calculated that the Labofina test involved dispersion of oil as droplets with diameters less than 190 um (depending on their density).

DISPERSANT-TO-OIL RATIO (DOR)

Results from laboratory tests indicate that measured values for dispersion are directly related to the dispersant-to-oil ratio (DOR). Figure 10 (from Rewick et al., 1981) presents dispersion curves for six dispersants (A through E) with No. 2 and No. 6 fuel oils. The amount of dispersant (in mL) was varied relative to a set volume of oil (100 cc or mL) for the tests. The data indicate that the quantity of dispersed oil is sensitive to DOR values less than 0.2 (i.e., 20 mL of dispersant), which is reflected by the steep slopes for percent dispersion in the curves. Similar effects of the DOR on dispersion values have been noted by other investigators (e.g., Mackay et al., 1978; Byford et al., 1984; Lehtinen and Vesala, 1984; Martinelli, 1984; Lewis et al., 1985; Fingas et al., 1990). Fingas et al. (1990) found that dispersion could approach 0% at DOR values of 1:40 to 1:60 (i.e., DOR approximately 0.02) for Alberta Sweet Mixed Blend crude oil and various dispersants. Byford et al. (1984) also have shown that increases in dispersion with increases in DOR are accompanied by production of smaller dispersed oil droplets (i.e., reduced volume median diameters or vmd), although the relationship between DOR and vmd is not linear. Frequently, a minimum of two or three DOR values are evaluated in laboratory testing. The data are then presented graphically to estimate the concentration of

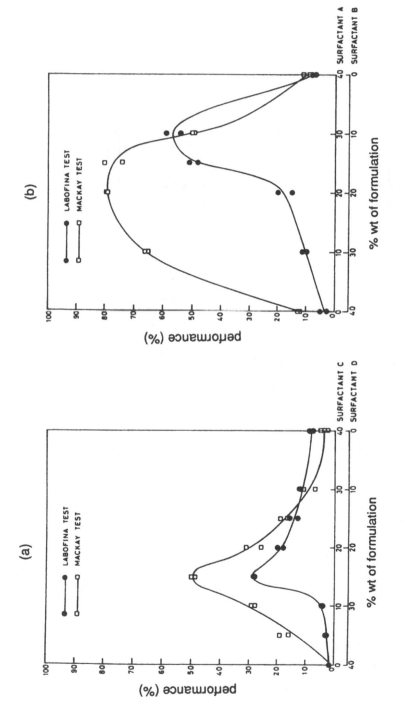

Figure 9. Effect of surfactant composition on Labofina and MNS (Mackay) dispersion performance for weathered Kuwait crude oil at 10ºC for (a) weathered Kuwait oil and surfactants C+D, (b) weathered Kuwait oil and surfactants C+D, (c) weathered Lago Medio oil and surfactants A+B, and (d) weathered Lago Medio oil and surfactants C+D. *Source*: Byford and Green, 1984. (Copyright American Society for Testing and Materials, 1984. Reprinted with permission.)

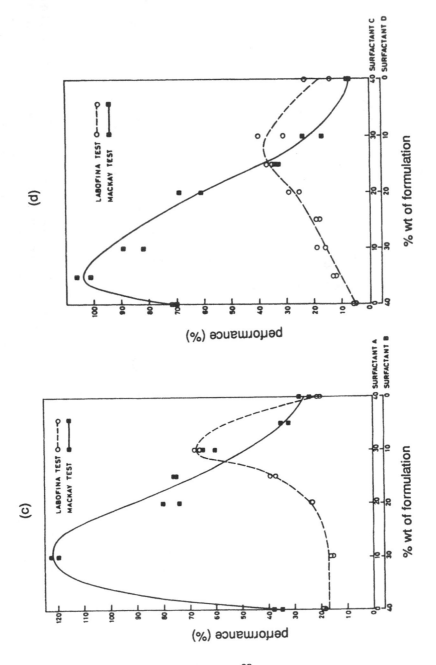

Figure 9. (continued)

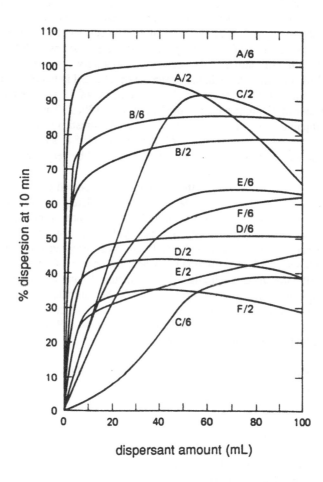

Figure 10. Dispersant performance curves for Nos. 2 and 6 fuel oils with different dispersant-to-oil ratios (DOR; varying dispersant amount with 100 mL of oil). *Source*: Rewick et al., 1981. (Copyright American Petroleum Institute, 1981. Reprinted with permission.)

dispersant that would be required to produce a target dispersion value.

OIL-TO-WATER RATIO (OWR)

The volume ratio of oil to water (OWR) used in laboratory tests can influence measured values for dispersion. For example, a strongly hydrophilic surfactant in a small volume of water will result in greater dispersion of oil than if a larger volume of water is used because the larger water volume will allow greater dissolution of the surfactant away from the oil-water interface. As a result, initial dispersions of oil might prove unstable in systems with low OWR values. As can be deduced from information in the section on laboratory test methods, the large capacity for dissolution losses of surfactants in the ocean is not reflected in many laboratory tests. In the Labofina test the ratio of oil to seawater (1:50) is a factor of 11 greater than in the MNS test (1:600) and a factor of 22 greater than in the EPA test (1:1300). Therefore, the dissolution and dilution of hydrophilic surfactants into the water can be drastically different in each of these tests, and may affect dispersion results. Mackay et al. (1984) suggest that under the most realistic application conditions at sea, the OWR normally will be very low.

Another aspect of the OWR is the possibility of dispersed oil droplet-droplet collisions and coalescence. Mackay et al. (1984) believe that it is unlikely that droplet-droplet collisions occur with sufficient frequency to cause appreciable coalescence in real open-ocean conditions. This is not the case, however, in laboratory tests with high oil-to-water ratios. In the latter situation, coalescence might occur and result in lower estimates for dispersion. This is particularly problematic with the Labofina test where the density of oil droplets can be estimated on the order of 10^8 drops/mL (assuming a drop diameter of 7 um and 100% dispersion). In the MNS test, the number density of oil droplets would be estimated as an order of magnitude smaller (i.e., 10^7 drops/mL). In open ocean conditions, oil concentrations in water rarely exceed 200 ppm (w:v). This results in a number density of 10^6 drops/mL, and the expected droplet collision frequency would be correspondingly lower.

Examinations of the effect of OWR on dispersion values in laboratory tests have been performed by Fingas et al. (1989b). Tests were conducted with three oils (Alberta Sweet Mixed Blend crude and two Beaufort Sea crudes identified as Adgo and Atkinson) and five dispersants (Corexit 9527, Enersperse 700, Corexit CRX-8, and two experimental formulations). Dispersant-to-oil ratios (DOR) were maintained constant at 1:25, while OWR values ranged from approximately 1:20 to 1:1,000,000. Experiments were performed with both the swirling flask and oscillating hoop tests. Results are shown graphically in Figure 11a and b. The data show that maximum dispersant performance occurred at OWR values of 1:500-1:600 and declined at higher oil concentrations between 1:20 and 1:500. Dispersion was essentially constant at OWR values between 1:600 and 1:1,000,000. Fingas et al. (1989b) postulate that the variations in dispersion at particular OWR values are due to different mechanisms affecting dispersant actions in specific OWR ranges. For example, at high oil-to-water ratios between 1:20 and 1:500, the concentrations of surfactant molecules are high enough to not only act like a detergent and remove oil from glass surfaces in the test flask but also form micelles that are comprised of only surfactant. The latter situation would occur at concentrations greater than

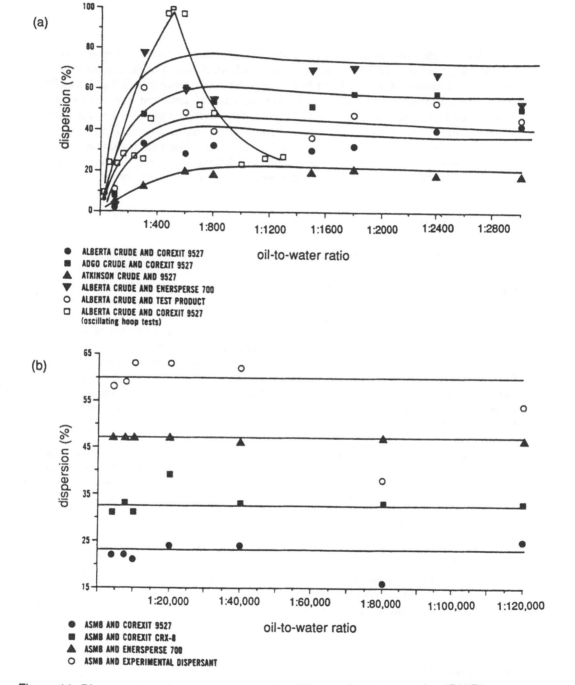

Figure 11. Dispersant performance curves with different oil-to-water ratios (OWR).
Source: Fingas et al., 1989b. (Copyright American Petroleum Institute, 1989. Reprinted with permission.)

the critical micelle concentration for the surfactant molecules. Surfactant molecules in the micelles would not be available to interact with the oil, which would result in a decrease in the dispersion performance for the oil. At low oil-to-water ratios between 1:500 and 1:1,000,000 and a constant premixed DOR of 1:25, the concentration of surfactant molecules in the test system is not high enough to result in the formation of the surfactant micelles (i.e., the surfactant concentration is below its Critical Micelle Concentration). Consequently, all available surfactant molecules react with the oil, which results in values for dispersant performance that are essentially constant over the OWR range of 1:600 to 1:1,000,000. At OWR values in the range of 1:500 to 1:600, both mechanisms are operating and chemical dispersion of the oil is maximized.

TEMPERATURE

Temperature can play an important role in reactions related to the dispersion of oil by chemical dispersants. For example, ethoxylated surfactants in commercial dispersants have higher water solubilities at lower temperatures, which can lead to temperature-dependent losses of surfactants from the oil-water interface. This also may lead to changes in the overall HLB value for the surfactant mixture in the oil, with coincident effects on the dispersion properties of the oil. More importantly, temperature will affect viscosities and pour points of both oils and dispersants. This can be critical at lower temperatures because penetration and mixing of dispersants with more viscous oil generally will be less effective. Roll-off losses of dispersants from an oil slick to the surrounding water also may be enhanced with more viscous oil. As for dispersant formulations, lower viscosities of the formulations (e.g., at higher temperatures) will generally result in production of smaller dispersant droplets for application purposes (e.g., Lindblom and Cashion, 1983; Gill, 1981 and 1984).

Wells and Harris (1979) studied effects of temperature (as well as salinity) on dispersion performance of Lago Medio crude oil with two dispersant formulations (Corexit 9527 and BP1100X). Tests were performed with the MNS test. Dispersion of oil was greater when both dispersants were warmed to 18°C as opposed to 1.5°C before dispersant applications in the test apparatus.

Lehtinen and Vesala (1984) evaluated the effect of water temperature (i.e., 4°, 12°, and 15°C) on dispersion of oil with three dispersants (i.e., one "marine" and two "low salinity" formulations) and a light Russian crude in both fresh and weathered states. The MNS test was used for all studies. Dispersion increased with increasing temperature for the "marine" and one of the "low salinity" dispersants. Temperature effects on the second "low salinity" dispersant were inconclusive.

Fingas et al. (1991b) studied the effect of temperature on dispersion of Alberta Sweet Mixed Blend (ASMB) crude oil and the dispersant Corexit 9527 in the Swirling Flask test. Dispersion increased with increasing temperature.

Mackay et al. (1980) conducted dispersion tests at varying temperatures and in the presence of ice with premixed solutions of Alberta medium gravity mixed sour blend crude oil (both fresh and evaporatively-weathered with a 25% loss in volume) and the dispersant Corexit 9527. Both the MNS test (using premixed oil-dispersant) and an oscillating hoop test were used for the studies. Results showed that dispersion increased with increasing temperature for both fresh and weathered oil. Increasing amounts of ice caused progressive decreases in levels of mixing energy at the air-water interface, with accompanying decreases in dispersion of the oil.

Byford et al. (1983) conducted laboratory tests for dispersant performance under low temperature and low energy conditions in an attempt to simulate circumstances that could be encountered in the Arctic. Tests were performed with the Labofina rotating flask test, seven dispersants (Arochem D-609, BP 1100 WB, Corexit 9527, Corexit 9550, Dispolene 34S, Finasol OSR-5, and an experimental formulation), and two oils (a 175°C topped Lago Medio residue and a North Slope crude that had lost approximately 11% of its initial weight during evaporative weathering). Decreases in temperature from 10°C to 0°C generally resulted in slightly higher values for dispersion, although this trend was not conclusive. In similar tests in Byford et al. (1984), results indicate that dispersion generally increased in tests performed at 20°C as opposed to 0°C, although trends were not always consistent for this pattern.

Byford et al. (1983) also performed studies to evaluate the effect of ice on dispersant performance. Tests were conducted in a 25-L tank with dispersant applied by an air-assisted spray applicator. The tank system was used for the ice studies because the Labofina procedure is not suited to assess dispersion performance in the presence of ice. Results from the tank tests generally showed higher values for dispersion in the presence of ice. Specifically, effects of the ice in the test tank included dampening of wave action as well as introduction of a pumping action to the oil between pieces of ice. The general trend of increasing dispersion in the presence of ice is in contrast with the findings of Mackay et al. (1980), although Cox and Schultz (1981) report no reduction in dispersion with up to 50% ice cover in oscillating hoop tests.

SALINITY

Salinity of receiving waters can impact dispersion of oil by chemical surfactants. Specifically, the intent of dispersant formulations for marine use is to provide maximum dispersion at normal seawater salinities. Mackay et al. (1984) note that higher salinities will increase dispersion by slightly hydrophilic dispersants by deterring migration of surfactant molecules into the water phase, which is equivalent to a salting-out effect for the surfactant from the saline medium. Such a situation will tend to promote association of surfactant molecules with oil at oil-water interfaces, which is important for lowering oil-water interfacial surface tensions in the oil-dispersant mixture. Salinity also may affect the water solubilities of dispersant formulations from the standpoint of influencing reactions related to the hydrophile-lipophile balance (HLB) of the dispersant mixture.

Wells and Harris (1979) evaluated the effect of salinity on the dispersion of Lago Medio crude oil with five dispersant formulations (Corexit 9527, Oilsperse 43, BP1100X, Drew O.S.E.

71, and Corexit 8666). Tests were performed with the MNS test. Dispersion was higher for four of the five dispersants in seawater as opposed to freshwater. The exception (Corexit 8666) produced no dispersion at either salinity tested (i.e., 0 and 29-31 ppt).

Byford et al. (1983) tested a number of dispersants under low salinity, low temperature, and low energy conditions to assess dispersant performance in conditions similar to those that could be encountered in the Arctic. Tests were performed with the Labofina rotating flask test, seven dispersants (Arochem D-609, BP 1100 WB, Corexit 9527, Corexit 9550, Dispolene 34S, Finasol OSR-5, and an experimental formulation), and two oils (a 175°C topped Lago Medio residue and a North Slope crude that had lost approximately 11% of its initial weight during evaporative weathering). Dispersion increased with increasing salinity for 5 of the 7 dispersants, although the increases did not always occur over the entire salinity range tested (i.e., 0 to 33 ppt). Only slight or no increase in dispersion with increasing salinity was observed for the remaining two dispersants. In summary, increasing salinity generally produced increasing dispersion, although the specific trends were dependent on the specific dispersant tested.

Lehtinen and Vesala (1984) evaluated the effect of salinity (i.e., 3, 7, and 12 ppt) on dispersant performance for three dispersants (i.e., one "marine" and two "low salinity" formulations) and a light Russian crude in both fresh and weathered states. The MNS test was used for all studies. Generally, dispersion increased with increasing salinity for the "marine" dispersant, whereas trends for the "low salinity" dispersants were less conclusive.

Fingas et al. (1991b) evaluated effects of salinity on dispersant performance for three crude oils (Alberta Sweet Mixed Blend or ASMB, Norman Wells, and Adgo) and three dispersants (Corexit 9527, Enersperse 700, and Citrikleen) in the Swirling Flask test. Maximum values for dispersion were obtained in the salinity range of 40-45 ppt. Increasing dispersion with salinity increases from 0 to 40 ppt imply that ionic interactions in the aqueous-oil system are important for dispersion with the dispersants evaluated. The results support the concept of a salting-out effect in which surfactant molecules are concentrated and stabilized in the oil phase (e.g., at the oil-water interface) as opposed to dissolving in the water phase. Reasons for declines in dispersion at salinities greater than 45 ppt are not entirely clear, but may be related to the fact that the main surfactants in the tested dispersants are nonionic and the HLB of these is strongly dependent on ionic strength.

Belk et al. (1989) evaluated dispersion of oil with one marine and two freshwater dispersant formulations in solutions containing not only different overall salinities but also different electrolyte compositions based on sodium, calcium, or magnesium ions. Tests were performed in the Labofina test apparatus with two oils: a Warren Spring Laboratory test oil (viscosity 2000 mPas at 10°C) and Prudhoe Bay crude. Results indicated that dispersion with all dispersants was lower at 0 ppt salinity, although exact trends differed for each dispersant as the salinity increased. As for specific electrolyte solutions, dispersion increased with increasing sodium ion concentrations but remained essentially constant or declined with increasing calcium and magnesium electrolyte concentrations for the marine dispersant. Dispersion with the two freshwater dispersants showed a smaller response with changes in the electrolyte type and concentration, although one of the freshwater dispersants showed slightly higher dispersion at lower calcium ion concentrations. The latter results imply that the electrolyte composition of an aqueous medium (as opposed to salinity alone) can affect dispersion for different dispersant formulations.

SAMPLING AND ANALYSIS METHOD

Numerous laboratory studies have included comparisons of dispersion-performance among different testing procedures. As already shown in Figure 8, Byford and Green (1984) found reasonably good agreement between results with the MNS and Labofina procedures. However, poor or no agreement frequently occurs between dispersion values with different procedures (e.g., Meeks, 1981; Rewick et al., 1981; Byford, 1982; Daling and Lichtenthaler, 1986/87; Gillot et al., 1986/87; Fingas et al., 1987b, 1989a; Daling et al., 1990b). The latter observation also is supported by laboratory studies that were conducted in conjunction with preparation of this book (Clayton et al., 1992). For example, measurements of dispersant performance were obtained with three testing procedures: the Revised Standard EPA test (10-minute and 2-hour samples), the Swirling Flask test (premixed, 1-drop, and 2-drop dispersant-to-oil additions), and the IFP-Dilution test (see Section 4 for descriptions of the procedures). Tests were performed with two oils (Prudhoe Bay crude and South Louisiana crude) and three commercial dispersant agents (Corexit 9527, Corexit CRX-8, and Enersperse 700). Controls with no dispersant agents also were included. Results of the dispersion measurements with the different testing procedures are illustrated in Figures 12 and 13 for Prudhoe Bay crude and South Louisiana crude, respectively.

Values for dispersion performance in Figures 12 and 13 for the test oils and dispersants generally follow trends of: EPA-10 minute > EPA-2 hour ~> Swirling Flask-premix > Swirling Flask-1 drop ~ Swirling Flask-2 drop. Results for the IFP-Dilution test generally vary between those for the three versions of the Swirling Flask procedure and are normally less than values from the Revised Standard EPA procedure. For the Swirling Flask procedure, dispersion performance is consistently higher with the premixed version of the test as opposed to the 1-drop or 2-drop versions. The 1- and 2-drop versions of the test, which involve dropwise additions of a dispersant to an oil slick, are intended to allow for an approximate estimation of the effect of dispersant-herding of oil on test results. As discussed in Section 3, herding occurs when dispersant droplets interact directly with the water surface. Surfactant molecules from a dispersant that are introduced onto the water surface have a driving force (or spreading pressure) to orient themselves at the interface based on their hydrophilic-lipophilic components. The result is that the surfactant molecules can push or herd the oil aside, which may lead to reductions in dispersion performance when the dispersant does not gain access to the oil. While herding also can occur during dispersant applications in the Revised Standard EPA and IFP-Dilution procedures, final results for these procedures do not appear to be substantially affected by herding (e.g., results with both of these procedures are often higher than those from the premixed version of the Swirling Flask procedure, especially for the Revised Standard EPA test). The lesser influence of herding on results from the EPA and IFP-Dilution procedures likely results from the dispersant-application protocols used in each procedure. For example,

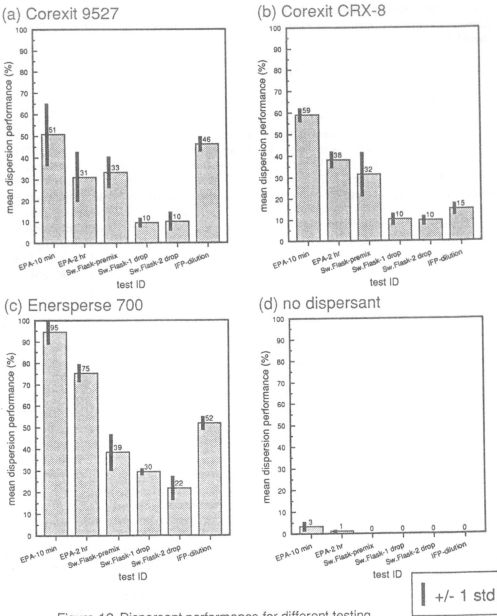

Figure 12. Dispersant performance for different testing procedures. Values are means and standard deviations (n=3 to 13). Oil = Prudhoe Bay crude.

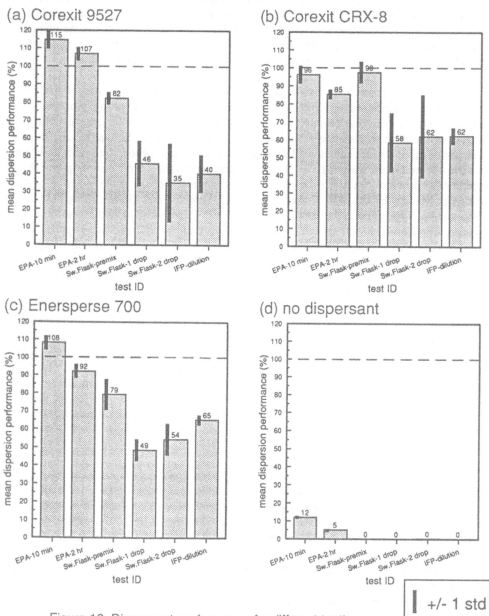

Figure 13. Dispersant performance for different testing
procedures. Values are means and standard deviations
(n=3 to 12). Oil = South Louisiana crude.

the initial oil slick is constrained in a containment ring during dispersant application in each of the procedures, which may minimize the ability of the dispersant to herd the oil aside and thereby reduce the interaction between the dispersant and the oil. Furthermore, the pressurized-spray application of seawater following dispersant application in the Revised Standard EPA procedure effectively returns any herded oil to the water's surface in the test tank.

At least a portion of the variability in experimental results among testing procedures can be attributed to variations in sampling approaches inherent to the testing procedures. For example, considerable variation exists in the time of sample collections for estimating dispersion performance relative to cessation of agitation in testing procedures. Subsurface concentrations of dispersed oil in a water column will depend on diameters and, hence, rise velocities of the oil droplets according to Stokes' Law. Consequently, incorporation of a settling-time into an experimental protocol can affect the measured concentrations of oil in the water column over time (e.g., Mackay et al., 1978; Byford et al., 1984; Lehtinen and Vesala, 1984; Fingas et al., 1989a and b). Typical times for sample collections after agitation is stopped in testing protocols include the following: MNS test (samples collected while agitation is still being applied after 10 minutes), Revised Standard EPA test (samples collected while agitation is still being applied after 10 and 120 minutes), oscillating hoop test (samples collected while agitation is still being applied), Warren Spring Laboratory rotating flask test (1 minute), and Environment Canada Swirling Flask test (10 minutes). Fingas et al. (1989b) studied the effect of settling-time (i.e., the time between cessation of agitation in the test flask and collection of a sample for oil analysis) on dispersion performance. Tests were performed with three procedures (the oscillating hoop, MNS, and swirling flask procedures), three dispersant formulations, and twelve oils. Samples for estimating dispersion performance (i.e., oil remaining in suspension in the water) were removed at selected time points after agitation had ceased for each of the test procedures. Results indicated that values for dispersed oil decreased with increasing settling-times, and the authors proposed that a standard settling-time of 10 minutes be applied to laboratory tests. Lehtinen and Vesala (1984) also found that measured values for dispersion in the MNS test decreased initially after agitation ceased, but then stabilized with settling-times greater than 10 minutes.

Fingas et al. (1989b) provide evidence that agreement between values for dispersion of oil in different test methods can be improved if the following procedures are incorporated into experimental protocols: (1) settling-times of approximately 10 minutes are adopted after agitation is stopped to allow for unstable dispersed oil droplets to rise to the surface, (2) the oil-to-water ratios are no more concentrated than 1:1000 (v:v), and (3) account is taken of the natural dispersibility for a given oil in the absence of chemical dispersant agents in a particular testing procedure. Fingas et al. performed experiments to evaluate dispersant performance with 16 types of oil, 3 commercial dispersant formulations, and 4 laboratory testing procedures (swirling flask, flowing cylinder, Labofina modified, and MNS modified). Modifications in the Labofina and MNS test procedures included premixing of dispersant with oil at a dispersant-to-oil ratio of 1:25 (v:v) and spectrophotometric detection of oil at wavelengths specified in Fingas et al. (1987b). Results of the studies with and without modifications to the four testing methods are graphically summarized in Figure 14a-c. Incorporation of the specified modifications into

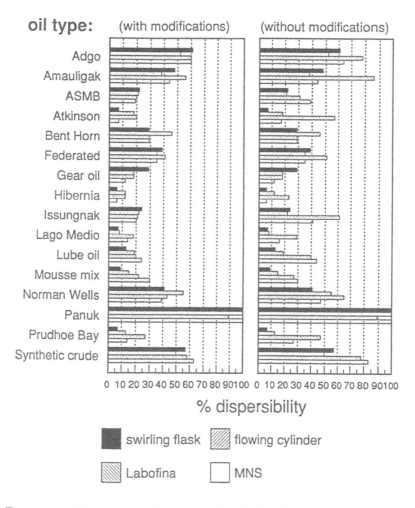

oil type:

Figure 14. Dispersant performance for 16 oils with four testing
methods. Data adapted from Fingas et al. (1989b).
(a) Corexit 9527 dispersant.

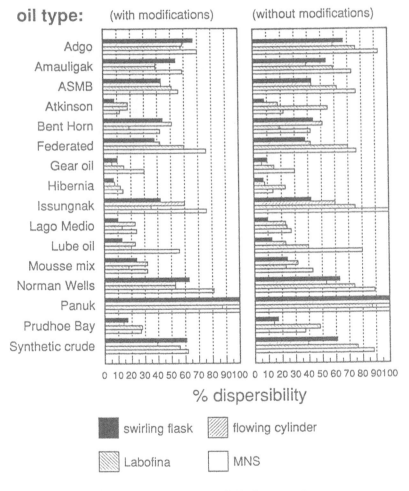

Figure 14. (continued) (b) Enersperse 700 dispersant.

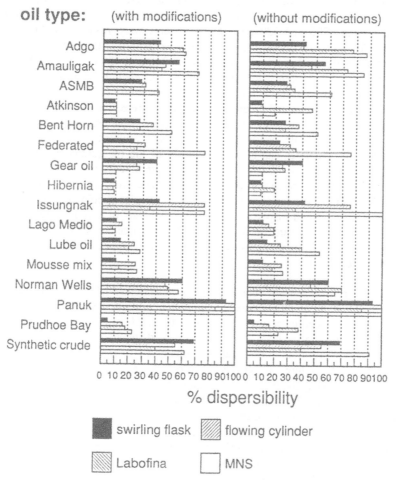

oil type: (with modifications) (without modifications)

% dispersibility

swirling flask   flowing cylinder

Labofina   MNS

Figure 14. (continued) (c) Corexit CRX-8 dispersant.

the protocols improved agreement between different test results for many of the oil types and dispersants evaluated.

In addition to experimental protocols, the analytical methods chosen to quantify amounts of dispersed oil in samples also is important for dispersion measurements. The most widely used methods for quantifying amounts of dispersed oil involve extraction of a given volume of seawater with a suitable solvent and quantitation by UV-visible spectrophotometric or (less frequently) gas chromatographic methods. Spectrophotometric determination of quantities of oil are performed at a variety of wavelengths (e.g., 580 nm in the standard protocols for the MNS, IFP-Dilution, and Labofina tests; 620 nm in the Revised Standard EPA test). Fingas et al. (1987b) evaluated detectability of oil in methylene chloride by spectrophotometry at UV-visible wavelengths between 200 and 720 nm. Measurements were made with three types of oil (Alberta Sweet Mixed Blend or ASMB, ASMB plus the dispersant Corexit 9527, and Issungnak crude). Analyses performed at wavelengths between 340 and 400 nm (i.e., 340, 370, and 400 nm, with the mean oil concentration for the three wavelength measurements being used) provided the best results in terms of consistency and repeatability for the oils tested. Analyses at wavelengths less than 340 nm and greater than 470 nm suffered from limitations including lack of linear response with concentration, poor reproducibility of values, and/or poor sensitivity in the spectrophotometer. Consequently, choice of analysis wavelengths for spectrophotometric measurements of oil quantities can affect the quality of the measurement values for oil and, hence, dispersion estimates. Fingas et al. (1990) also note that the presence of water and some dispersants in methylene chloride can affect the absorbance readings in a spectrophotometer. Therefore, calibration curves for quantitation of oil in samples should be developed by adding specified amounts of oil and dispersant, if used, to volumes of water equivalent to those in samples, extracting the standard oil-water solutions in a manner identical to a sample, and performing the spectrophotometric measurements on the standard-extracts to generate the calibration curve for the oil.

If values for dispersion are obtained with good analytical laboratory measurements, then calculations might be made to estimate the amount of a particular dispersant that would be required to achieve a particular dispersion value (e.g., 50%). As noted above, however, different laboratory testing methods can yield different dispersion values for the same oil and dispersant types if appropriate consideration is not given to the experimental and analytical details of the testing procedure. To reduce effects of as many experimental variables as possible (e.g., OWR, DOR, mixing energy, etc.), Rewick et al. (1980, 1981, 1984) proposed methods related to changes caused by dispersants in oil-water interfacial surface tensions to estimate dispersant performance. Initially, measurements of interfacial surface tension were made with a surface tensiometer and du Nouy ring (Rewick et al., 1980). However, this approach suffered from limitations including not only difficulties dealing with viscous oils in the du Nouy ring/tensiometer apparatus but also the fact that the dispersant and oil had to be premixed. Consequently, the Drop-Weight test was developed to measure Critical Micelle Concentrations (CMC) for dispersant-oil mixes (Rewick et al., 1981, 1984). Data generated with the Drop-Weight procedure can be used to estimate and rank relative dispersant performance and effectiveness. Rewick et al. (1981, 1984) admit, however, that results of the drop-weight test cannot be compared directly with results from other tests because no mixing energy is required and results are not directly interpretable in terms of quantifying either actual amounts of dispersed oil produced or the dispersant-to-oil ratios that are critical for the dispersion reaction. Nevertheless, the method does address a fundamental physical property that applies to dispersant-oil interactions: the lowering of the oil-water interfacial surface tension and, hence, the true effectiveness of specific chemical dispersant agents. As such, the measurements expressed in terms of critical micelle concentration are independent of variables such as mixing energy and wall effects that are difficult to control in other testing methods.

Another approach for assessing dispersant performance has been to examine sizes of chemically dispersed oil droplets as functions of not only dispersant type but also variables such as temperature and energy input (e.g., Byford et al., 1984; Lewis et al., 1985). In the latter studies, size distributions were determined for oil droplets produced with different testing procedures (i.e., the MNS, Labofina, and oscillating hoop tests). Droplet sizes were determined with a Malvern 2600D particle size analyzer, which consists of a low powered helium/neon laser and a receiver unit. Several beams of light are passed through a sample cell and diffraction caused by particles in the fluid is measured by a series of concentric photosensitive rings that measure the amount of light diffracted in a range of angles from the original central path. By measuring the amount of light in each detector, the number of particles in certain size ranges can be determined. Normally the Malvern system can determine particles between 0.5 and 560 um, although use of alternate lens and base units can extend the range up to 1800 um. Data generated for dispersed oil droplet sizes in the studies of Byford et al. and Lewis et al. suggest that the lack of agreement among dispersion values with different laboratory test methods might be explained by differences in mixing energies and sampling methods inherent to a particular procedure (e.g., higher mixing energies produce smaller, more stable dispersed oil droplets and application of sufficient settling-times can lead to disappearance of larger, less stable oil droplets from sampled solutions). Increases in the dispersant-to-oil ratio also produced smaller dispersed oil droplets, although this trend was dependent on the dispersant used. Overall results indicated that a wide size range of oil droplets could be dispersed into seawater by using different laboratory test methods. The data also suggest that sizes of dispersed oil droplets could be important for estimating performance of dispersant agents.

SECTION 4

LABORATORY TESTING OF DISPERSANT PERFORMANCE

The general purpose and objectives of laboratory testing are the following:

- Screen dispersant types as a prelude to recommending their use in spills or more expensive studies such as field testing (e.g., Meeks, 1981; Nichols and Parker, 1985), or determine limitations or restrictions for use of specific dispersant formulations. It must be emphasized, however, that it is unlikely that any one test will give results that are unequivocal as far as product performance is concerned because of the many variables affecting test results. Nevertheless, at a minimum laboratory tests can be used to provide information for investigated products, such that products demonstrating poor dispersion performance in the laboratory could be eliminated from further testing.

- Test dispersion performance for a variety of dispersants to rank relative performance of the various products (e.g., Mackay and Szeto, 1981; Meeks, 1981; Mackay et al., 1984; Martinelli, 1984; Rewick et al., 1981, 1984; Canevari, 1985, 1987; Desmarquest et al., 1985; Fingas et al., 1989a and b, 1991a and b). This requires that the testing procedure provide results that are sufficiently precise and reproducible to distinguish among performance values for different dispersant agents.

- Test the performance of dispersants under carefully controlled laboratory conditions to assess the role of oil type, weathering state, oil-to-water ratio, dispersant-to-oil ratio, mixing energy, salinity, temperature, and dispersant application methods (e.g., Mackay et al., 1984; Byford et al., 1983; Lehtinen and Vesala, 1984; EPA, 1984; Payne et al., 1985; Buist and Ross, 1986/87; Fingas et al., 1987a and b, 1989a and b, 1991a and b; Daling, 1988; Daling et al., 1990a).

- Provide data that can be used in conjunction with other information by On-Scene Coordinators for emergency response and contingency planning at real spills, stockpiling of specific dispersants for particular environments and oil types, and deciding whether or not particular dispersants should be used (e.g., Mackay et al., 1983).

- Generate data to validate and improve mathematical modeling efforts for predicting dispersant performance (e.g., Mackay, 1985; Daling et al., 1990b).

- Estimate concentrations for dispersants that might be more appropriate for toxicity testing (e.g., Mackay and Wells, 1983; Bocard et al., 1984; Anderson et al., 1985; Wells, 1985; Franklin and Lloyd, 1986/87).

A variety of laboratory testing methods have been used in attempts to evaluate dispersant performance. In general, laboratory tests can be placed into four categories: (1) tank tests with

54

water volumes ranging from 1 to 150 L, (2) shake/flask tests that are conducted on a relatively smaller scale and require less sophisticated laboratory equipment, (3) interfacial surface tension tests that measure properties of the treated oil instead of dispersant performance directly, and (4) flume tests using flowing water systems with the capacity for breaking/nonbreaking waves to generate energy regimes that can more closely simulate real-world conditions in large water bodies (e.g., oceans and coastal bays). Each type of test uses a general approach of (1) establishing an oil slick on water, (2) applying dispersant to the slick, (3) applying energy to the oil-dispersant-water system, and (4) measuring the amount of oil dispersed into the water (Canevari, 1985).

Generally, dispersant performance in tests is determined by one of the following methods:

- Water column concentrations of dispersed oil are determined by either solvent extraction and spectrophotometric determinations or simple visual observations.

- Percent oil remaining in the surface slick is estimated (e.g., Nichols and Parker, 1985).

- Dispersed oil droplet sizes are determined (Byford et al., 1984; Lewis et al., 1985; Daling et al., 1990b).

- Interfacial surface tension measurements are determined and extrapolated to dispersant effectiveness (e.g., Mackay and Hossain, 1982; Rewick et al., 1981, 1984).

- Stable dispersed oil droplet quantities and properties are evaluated as functions of time and mixing energy in both static and dynamic systems (e.g., Mackay et al., 1984; EPA, 1984; Fingas et al., 1989a and b, 1991a and b).

LABORATORY TESTING METHODS

At least 35 methods for testing dispersant performance have been developed (e.g., as reviewed in Meeks, 1981; Rewick et al., 1981; Mackay et al., 1984; CONCAWE, 1986; Woodward-Clyde Consultants and SRI International, 1987; Fingas et al., 1989a and b). A number of significant differences are inherent to the various methods. For example, different methods for adding the dispersant to oil have included premixing of dispersant with oil, slowly pouring dispersant onto the oil, spraying the oil surface with a fine mist of either neat dispersant or dispersant premixed with seawater, and pouring the dispersant in the water prior to adding the oil. Premixing of dispersant with oil or adding a dispersant to seawater in the absence of oil are not completely realistic of real-world spill situations. Test-specific variations in the ratio of the oil-to-water volumes can have significant impact on not only the relative performance of dispersants (e.g., hydrophilic versus lipophilic) but also the magnitude of wall-effects in test containers. A variety of approaches have been used to provide mixing energy to test systems including a circulating pump and spray hose system (Revised Standard EPA test), a high velocity air stream that produces small waves of 1-6 cm height (MNS test), raising and lowering of a metal hoop (IFP-Dilution and oscillating hoop tests), rotating separatory funnels (Labofina/Warren Spring Laboratory tests), shaking flasks on a shaker table (Environment Canada/Swirling Flask test), and vertically flowing water (Environment Canada/flowing cylinder test). Additional means to generate mixing can include high speed propellers, oscillating paddles, and air bubbles. Samples are withdrawn for analyses after mixing is stopped in the Environment Canada/Swirling Flask and Labofina/Warren Spring Laboratory tests; whereas, sampling is generally conducted under dynamic mixing conditions in the Revised Standard EPA and MNS tests. Samples from the Revised Standard EPA test are withdrawn from the bottom of a 130 L tank after passing through a circulation pump and connecting plastic tubing, while sampling for the MNS test is conducted through a tube positioned in the center of the apparatus. In the Labofina/Warren Spring Laboratory test, a sample is withdrawn from the bottom of the separatory funnel after a typical settling period of 1 minute. In the Swirling Flask test, the sample is withdrawn from the bottom of the flask after a settling-time of 10 minutes. In summary, the wide variety of test conditions can make comparison of results among different methods quite problematic. More detailed descriptions of some of the more commonly used and/or unique testing procedures are considered in the following sections.

Tank Tests

MNS (Mackay/Nadeau/Steelman) Test--

The MNS test has been described in Mackay and Szeto (1981), Mackay et al. (1984), and Anonymous (1984). The standard test uses a glass tank placed in a temperature-controlled water bath (Figure 15). The tank has an external diameter of 310 mm and a height of 310 mm. A Plexiglas lid, attached to the top of the tank, has separate ports for (1) collecting oil, (2) adding oil to a containment ring, (3) inserting a thermometer, (4) adding dispersant, (5) allowing air flow into the tank, and (6) air exhaust. Air flow into the tank is used to generate mixing energy from surface waves with heights of 1-6 cm. A volume of seawater (e.g., 6 L) is added to the tank. The system is maintained at specified temperatures such as 0-2°C. Oil (e.g., 10 mL) and then undiluted dispersant are added inside the containment ring. If premixed dispersant-seawater mixtures are used, they may be added with a spray system fabricated from 10 mL plastic syringes attached back-to-back such that the mixture of dispersant and seawater can be forced by air pressure through an atomizing nozzle onto the oil. Inflow of air to the test system during a test is adjusted to maintain waves at specified heights (e.g., greater than 3 cm). The level of mixing energy in the tank cannot be directly quantified. However, the agitation can be reproduced by monitoring the pressure drop in a pressure plate manometer that is upstream of the air being introduced into the system. The air can be precooled by passage through an ice-water bath. Whether a dispersant is added by pipette or from the spray nozzle, the dispersant drops are applied to the oil from a uniform height of approximately 10 cm over a 10-30 second period. A 1-minute soak-period follows to allow the dispersant to penetrate into the oil, after which the containment ring is removed and air flow is continued. After 10 minutes of agitation, suction is applied to the sample-collection tube and a 500-mL sample of water is siphoned into a 1-L separatory funnel. Sufficient sample (approximately 50 mL) is allowed to pass through the sampling tube to clear the void volume in the tube before the actual sample is collected. In

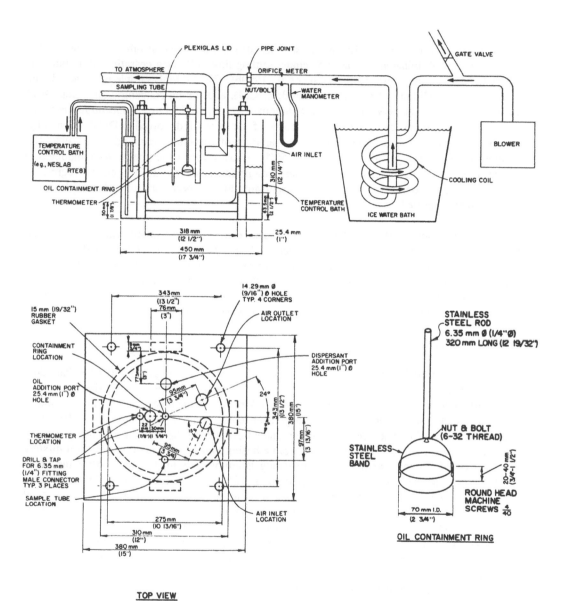

Figure 15. MNS test apparatus. *Source*: Mackay and Szeto, 1981. (Copyright
American Petroleum Institute, 1981. Reprinted with permission.)

addition to the 10-minute sampling-event, samples also can be drawn at other times (e.g., after only 5 minutes of agitation or after a 5-minute settling-time following cessation of agitation in the tank; Mackay and Szeto, 1981; Daling et al., 1990b). The 500-mL sample is extracted with three 20-40 mL aliquots of methylene chloride (DCM), and the oil concentration in the DCM extract is determined by UV-visible spectrophotometry (e.g., at 580 nm). Concentrations of the test oil are determined against a calibration curve of the same oil dissolved in DCM. Generally, results are presented as a curve showing the amount of dispersant required to disperse 100 volumes of oil with a performance-value of 50%. Usually four experiments are conducted in duplicate, spanning the range of 30-70% dispersion performance in order to obtain a regression line. The 95% confidence intervals in dispersant dosage necessary to achieve the 50% dispersion are reported.

During conduct of the MNS test, wave dampening has been noted with some oils and dispersants (e.g., Mackay et al., 1984). Such a phenomenon can affect measured values for dispersion because input of turbulent energy to the system is reduced. Compensation for wave dampening can be done by applying a higher air flow rate to the test apparatus. P.S. Daling (personal communication) has determined that reducing the agitation period from 10 to 5 minutes in the MNS test apparatus also will eliminate wave dampening.

Revised Standard EPA Test--

The Revised Standard EPA test, described in the Federal Register (40 CFR Part 300; EPA 1984), utilizes a stainless steel tank containing 130 liters of 25 ppt seawater (Figure 16). A measured weight of oil (ca. 100 mL in volume) is added to a containment ring held in the middle of the test tank. A measured weight of dispersant (corresponding to volumes of 3, 10, or 25 mL) is then added in a fine stream from a graduated cylinder to the oil over a 1-minute period. After dispersant is added, water is sprayed from a hosing system onto the dispersant-oil mixture for a period of 1 minute as the containment ring is removed from the tank. After the initial spray-mixing, the tank water is recirculated by a 1/35 HP centrifugal pump through plastic tubing from the top to the bottom of the tank. Samples for dispersed oil analysis are removed from the bottom of the tank through the centrifugal pump/plastic tubing bypass after periods of 10 minutes and 2 hours of recirculation. Sample volumes of 500 mL are extracted with DCM or chloroform for oil analysis. Concentrations of oil in the extracts are analyzed by spectrophotometric methods at 620 nm. Tests with the EPA tank are conducted in triplicate. Replicate tests yielding values that vary by greater than 8% from the mean are repeated. Dispersant dosages causing 50% dispersion at 10 minutes and 25% dispersion at two hours are estimated.

Two studies have proposed modifications to the above EPA test in an effort to provide greater repeatability or reproducibility and environmental relevance to test results. Woodward-Clyde Consultants and SRI International (1987) retain the Revised Standard EPA test tank but incorporate changes to the overall design of the apparatus to address three problematic aspects of the test: (1) the need to apply energy to mix a dispersant with oil in a more reproducible manner, (2) development of a mixing-energy source for simulating wave and tidal action

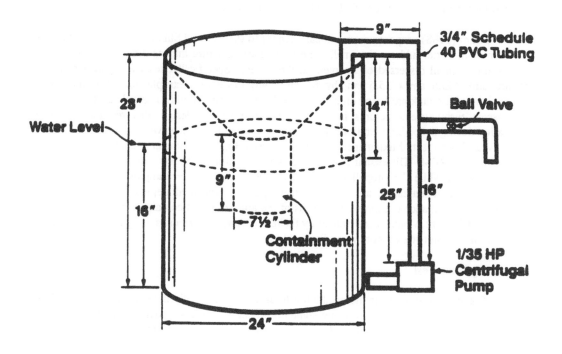

Figure 16. Revised Standard EPA dispersant test apparatus.

following application and mixing of the dispersant with the oil, and (3) development of a sampling device that would eliminate the circulation pump and connecting plastic tubing. Proposed changes include the following. Energy for mixing of a dispersant with oil on the water's surface is supplied by a standard, full cone spray nozzle, which allows spray to cover the entire water's surface including the oil-dispersant containment ring. The circulating pump system is removed from the tank design, and the energy source for simulating wave and tidal action is supplied by a low-shear paddle stirrer that is rotated at a specified rate. In the absence of the circulating pump and tubing system, samples are withdrawn from the test tank from ports welded into the tank wall at specified depths below the water's surface.

Shum (1988) proposes more radical changes to the Revised Standard EPA test design in an attempt to better address the small-scale turbulence structure that controls the dynamics of formation of small dispersed oil droplets in the ocean. Specifically, tests are performed in a square tank (35.6 cm x 35.6 cm, by 61 cm high) made of clear plastic. The horizontal dimensions of the tank are based on sizes of the largest turbulent eddies calculated to occur for small-scale turbulence structures in the ocean. The square design also minimizes vortex formation during propeller operation. The tank contains 38 liters of seawater. A 40-mL volume of oil is used for a test. Mixing energy is provided by a 30.5-cm diameter propeller mixer that is rotated at 210 RPM. Dispersant is premixed with the oil and applied with a syringe onto the water's surface while the propeller mixer is in operation. Water samples for dispersed oil content and characterization are collected from ports at specified depths in the tank walls. Dispersant performance is expressed in terms of the amount of oil contained in droplets smaller than a specified size, which is determined by the relationship between sampling time and depth in the tank and droplet sizes according to Stokes' Law.

Oscillating Hoop Test--

The general design for the oscillating hoop test was initially developed by D. Mackay. As described in Fingas et al. (1989a), no standard procedure has been adapted or defined for the test, although a number of slightly different protocols have been used. However, the following is the general approach taken for oscillating hoop tests. The physical components for the test apparatus consist of a cylindrical container (e.g., 50-L capacity) and a metal hoop with a diameter slightly less than that of the container. A volume of water (e.g., 35-L) is placed in the cylindrical vessel. The hoop is raised and lowered beneath the surface of the water in the container at a specified oscillation rate (e.g., 60-150 RPM) to generate inwardly moving, concentric waves. The oscillating hoop design is intended to maintain the oil on the surface of the water toward the center of the tank by the inward movement of concentric waves, which should minimize wall-effects such as losses of oil onto the vessel walls during the agitation period. The vessel usually has a sampling port at or near its bottom through which samples can be withdrawn. For tests designed to evaluate dispersant performance, the following is a typical protocol: the hoop oscillation is started, a volume of oil is placed in the center of the water's surface, a volume of dispersant is added to the oil slick, and agitation is continued for a specified period of time before a sample is taken from a collection port. Spectrophotometric analyses of solvent extracts of water samples are normally performed to estimate oil content in samples.

IFP-Dilution Test--

The IFP-dilution test and apparatus is described in Desmarquest et al. (1985), Bardot et al. (1984), Bocard et al. (1984), Daling et al. (1990b), and CEDRE (undated). A schematic diagram of the apparatus is shown in Figure 17. The physical apparatus for the test involves a cylindrical glass container for holding a test solution and an oscillating hoop that fits inside the container. The glass container has two ports: (1) an inlet port located just below the experimental water level and (2) an outlet port that is located near the bottom of the vessel and contains an overflow arm extending upward to determine the depth of the test solution in the container. Clean seawater is introduced by a peristaltic pump into the glass container through the inlet port. Overflow water (containing oil droplets) leaves the container through the exit port and is collected in a flask. The oscillating hoop (external diameter 155 mm, internal diameter 125 mm) is suspended 20-35 mm beneath the water's surface and moved up and down with a 15 mm vertical path by an electromagnet controlled by an electronic timer. The frequency of the oscillation can be varied in the range of 6.66-20 cycles/minute (e.g., 15 cycles/minute in Daling et al., 1990b). For tests designed to evaluate dispersant performance, the following experimental protocol is followed: the glass container is filled with a known volume of seawater (e.g., specified volumes ranging from 4-5 L in different references; Desmarquest et al., 1985; Bardot et al., 1984; Bocard et al., 1984; Daling et al., 1990b; CEDRE, undated), a specified amount of oil (e.g., 4 g or 5 mL) is poured onto the water surface inside a 6-10 cm diameter vertical ring, dispersant is added onto the surface of the oil, the oscillating hoop is started, and water flow through the peristaltic pump is started at a specified flow rate to produce a dilution or turnover rate of 0.5 volumes/hour. Outflow water is collected for specified periods of time (e.g., 0-60 minutes in Daling et al., 1990b, and CEDRE, undated; 0-30, 30-60, and 60-120 minutes in Desmarquest et al., 1985, Bardot et al., 1984, and Bocard et al., 1984). Water samples are extracted with an appropriate solvent (e.g., DCM) and analyzed spectrophotometrically at 580 nm for oil content. The oil content in samples from protocols with multiple sampling events follows the equation:

$$x = x_0 e^{-Dt} \tag{3}$$

where x = final oil concentration at time t,
$\quad x_0$ = initial oil concentration at the start of the sampling event, and
$\quad D$ = dilution rate (i.e., flow rate divided by volume of water in tank).

The percentage of washed-out oil (P) at time t is:

$$P = 100(1-x/x_0) = 100(1-e^{-Dt}). \tag{4}$$

Dispersion performance due to a chemical dispersant is determined from the equation:

$$E = 100[(P_d-P_c)/P_d] \tag{5}$$

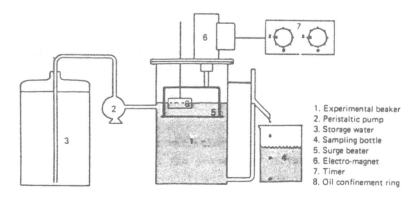

1. Experimental beaker
2. Peristaltic pump
3. Storage water
4. Sampling bottle
5. Surge beater
6. Electro-magnet
7. Timer
8. Oil confinement ring

Figure 17. IFP-Dilution test apparatus. *Source*: Daling et al., 1990b. (Copyright Oil & Chemical Pollution Journal, Elsevier Science Publishers Ltd., 1990. Reprinted with permission.)

where E = dispersant performance (%),

      P_d = percentage of washed-out oil at time t in solution with dispersant, and

      P_c = percentage of washed-out oil at time t in control solution without dispersant.

Flowing Cylinder Test (Environment Canada)--

The flowing cylinder test was developed by Environment Canada and is described in Fingas et al. (1989b). A schematic representation of the apparatus is shown in Figure 18. The basic operating principle in the apparatus is that water is continuously removed from the bottom of a cylinder and replaced at the top of the cylinder. The vertical circulation draws dispersed oil droplets into the water column and ultimately into a filter that removes the oil. Water without oil droplets returns to the top of the cylinder. The vertical descent of the water through the cylinder provides the energy for dispersion of the oil droplets. The length of the cylinder is sufficiently long so that only small oil droplets (e.g., 1-30 um in diameter) reach the exit port from the cylinder. Larger oil droplets return to the surface oil slick due to buoyancy forces. The filter holder contains the following three-stage filter sequence: (1) 5 um pore size filter, (2) 0.22 um pore size filter, and (3) a backup filter pad. For tests designed to evaluate dispersant performance, the following is a typical experimental protocol: the cylinder is filled with 1000 mL of seawater, the peristaltic pump is started at a specified flow rate (e.g., 100 mL/minute), a premixed volume of oil+dispersant is carefully placed in the center of the water surface, and the apparatus is allowed to operate for a specified period of time (e.g., 10 circulations lasting 100 minutes). At the end of the experiment, the filters are removed, extracted with a solvent (e.g., methylene chloride), and the extract is analyzed spectrophotometrically for oil with procedures described in Fingas et al. (1987b).

Shake/Flask Tests

Rotating Flask Test (Labofina/Warren Spring Laboratory)--

The Warren Spring Laboratory test is a modification of the Labofina test (Martinelli, 1984 and references therein). Both tests utilize 250-mL separatory funnels in a design apparatus shown in Figure 19. The shape of the separatory funnel is known to be important for obtaining precise and reproducible results of dispersion performance. Dimensions of the flask used at the Warren Spring Laboratory are shown in Figure 19. For a test, 250 mL of synthetic seawater are added to a separatory funnel, followed by 5 mL of oil. A 0.2-mL volume of dispersant is then added drop-wise to the oil surface. The flask is stoppered and secured into a holder that is mechanically rotated in the vertical plane through a 360° angle at 33 ± 1 RPM for 2 minutes. The flask is then allowed to stand for 1 minute after rotation is stopped, at which time 50 mL of water are removed through the stopcock of the flask into a graduated cylinder. The test is performed at a specified temperature (e.g., 10°C). The water sample is extracted with solvent (e.g., chloroform or methylene chloride), which is then analyzed for oil content by spectrophotometry at 580 nm.

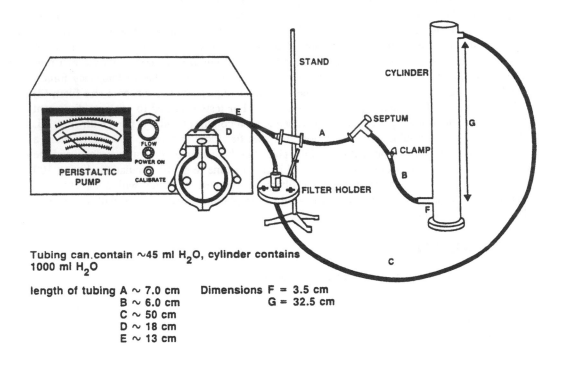

Tubing can contain ~45 ml H₂O, cylinder contains
1000 ml H₂O

length of tubing A ~ 7.0 cm   Dimensions F = 3.5 cm
              B ~ 6.0 cm   G = 32.5 cm
              C ~ 50 cm
              D ~ 18 cm
              E ~ 13 cm

Figure 18. Flowing cylinder test apparatus. *Source*: Fingas et al., 1989b. (Copyright
        American Petroleum Institute; 1989. Reprinted with permission.)

(a)

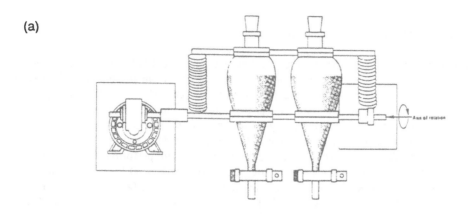

(b)

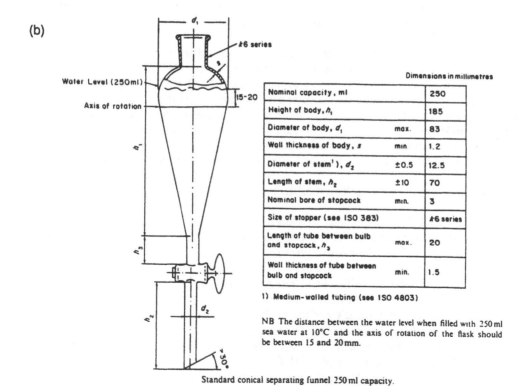

		Dimensions in millimetres	
Nominal capacity, ml			250
Height of body, h_1			185
Diameter of body, d_1	max.		83
Wall thickness of body, s	min.		1.2
Diameter of stem[1]), d_2	±0.5		12.5
Length of stem, h_2	±10		70
Nominal bore of stopcock	min.		3
Size of stopper (see ISO 383)			/6 series
Length of tube between bulb and stopcock, h_3	max.		20
Wall thickness of tube between bulb and stopcock	min.		1.5

1) Medium-walled tubing (see ISO 4803)

NB The distance between the water level when filled with 250 ml sea water at 10°C and the axis of rotation of the flask should be between 15 and 20 mm.

Standard conical separating funnel 250 ml capacity.

Figure 19. Warren Spring Laboratory rotating flask test apparatus. (a) *Source*: Daling et al., 1990b. (Copyright Oil & Chemical Pollution Journal, Elsevier Science Publishers Ltd., 1990. Reprinted with permission.) (b) from M. Webb (personal communication).

Swirling Flask Test (Environment Canada)--

The Swirling Flask test was developed in Environment Canada's laboratory to provide a relatively rapid and simple testing procedure for evaluating dispersant performance (Fingas et al., 1987a). The container for the test is a 125-mL Erlenmeyer flask to which a side spout is added to remove subsurface sample volumes near the bottom of the flask without disturbing the surface oil layer (Figure 20). Mixing energy is provided by placing the flask on a standard shaker table with a specified shaking rate (e.g., 150 RPM) to induce a swirling motion in the liquid contents of the flask. For tests designed to evaluate dispersant performance, the following procedure is observed: 120 mL of seawater are added to the flask, 0.1 mL of a premixed dispersant-oil solution (DOR = 1:25) is carefully added onto the surface of the water, and agitation is performed for 20 minutes. The flask is then immediately removed from the shaker table and kept stationary for 10 minutes, after which a 30-mL water sample is collected through the subsurface sampling port of the side spout in the flask. The sample is extracted with DCM and analyzed spectrophotometrically (i.e., at wavelengths of 340, 370, and 400 nm) for oil content by procedures described in Fingas et al. (1987b).

Subsequent adaptations to the Swirling Flask test include protocols in which 0.1 mL of oil without dispersant is added to the surface of the water in the flask, followed by one drop (10 uL) or 2 drops (5 uL each) of dispersant to the oil slick (DOR = 1:10; Fingas et al., 1990; M. Fingas, personal communication). All other experimental procedures in the one-drop and two-drop versions of the test are identical to those with the premixed dispersant-oil solutions. The one-drop and two-drop versions of the test are intended to assess effects of herding of oil by the dispersant (i.e., surfactant molecules orient at the air-water interface, which can result in pushing aside of the oil and a reduction in the interaction between surfactant molecules and the oil).

Exxon Dispersant Effectiveness Test (EXDET)--

The EXDET procedure was developed at Exxon Chemical Company (Becker et al., 1991) and is a modification of the Rotating Flask Test (Labofina/Warren Spring Laboratory). The test involves addition of 1 mL of a premixed oil-dispersant solution to 250 mLs of seawater in a 250-mL separatory funnel. The funnel is shaken in a wrist-action shaker for 15 minutes at a rate determined to yield a specific uptake rate for oxygen in the seawater (i.e., oxygen level in nitrogen-purged water increases from 1 to 3 ppm over a 4-minute period). A standing 1-inch wave is produced at the water's surface in the test vessel with the specified shaker rate. Calibration of the shaker rate for the specified oxygen-uptake rate is designed to standardize the agitation imparted to the seawater in the test vessel. After 15 minutes of shaking, a square of sorbent pad (1.5 inches x 1.5 inches; 3M Company) is added to the separatory funnel and shaking is continued for an additional 5 minutes. The purpose of the sorbent pad is to adsorb non-chemically-dispersed oil in the test solution. At the conclusion of all shaking, the water is drained from the separatory funnel and extracted with solvent (e.g., chloroform). Chloroform is also used to extract all remaining oil from the sorbent pad and walls inside the separatory funnel. The separate water and sorbent pad-separatory funnel extracts are analyzed for oil content by spectrophotometry at 460 nm. Results from the test are designed to (1) yield a mass balance for

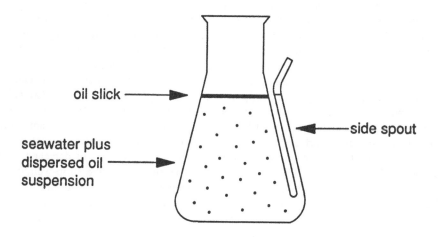

oil slick

side spout

seawater plus
dispersed oil
suspension

Figure 20. Swirling flask apparatus (modified 125-mL
              Erlenmeyer flask).

oil in the testing vessel (i.e., total oil is equal to that in the water and sorbent pad-separatory funnel fractions) and (2) allow for estimation of oil in chemically-dispersed (i.e., water) and non-chemically-dispersed (i.e., sorbent pad-separatory funnel) fractions.

Interfacial Surface Tension Tests

Drop-Weight Test--

The drop-weight test was adapted by investigators at SRI and is described in Rewick et al. (1981, 1984). The test is intended to estimate dispersant effectiveness or capability based on changes in oil-water interfacial surface tension that occur at oil-water interfaces of oil drops. A diagram of the testing apparatus used for drop-weight measurements is presented in Figure 21. In a test, a premixed dispersant-seawater solution is added to a 20-cc serum vial. Oil is added to a syringe, which is then weighed. The syringe is attached to the serum vial and an oil drop is forced under pressure in the syringe to detach from the U-shaped capillary tube into the dispersant-seawater solution. The syringe needle must be placed at a constant position in the dispersant-seawater test solution to equalize buoyancy effects, and oil drops must be allowed to grow slowly before detachment. Following detachment of one or more drops, the syringe is weighed to determine the drop weight(s) of oil released from the syringe. The weight of the detached drop of oil is plotted against the concentration of dispersant in the seawater medium, as shown in Figure 22. Data points on the plot are needed at approximately 8 concentrations of dispersant ranging from 0 to 100 ppm. The data points generate a Critical Micelle Concentration (CMC) curve as shown in the latter figure, with the CMC being the point of intersection of two straight lines fitted by linear regression for a given dispersant formulation. Reductions in oil-water interfacial surface tension that result from dispersant interactions at the oil drop-water interface are proportional to differences in the weights of detached oil droplets calculated at 0 ppm and the CMC in the plot. The success of the drop-weight technique depends on the oil drop in the aqueous solution reaching an equilibrium surface coverage with surfactant molecules before detachment from the syringe needle. Evaluations of dispersant effectiveness from the information contained in the plot are based on values for (1) the actual CMC concentration, (2) the initial slope of the CMC curve ($slope_i$), and (3) the interfacial tension reduction (Δwt), assuming that a given dispersant agent (1) reaches full surface coverage at the oil-water interface at the lowest concentration, (2) promotes the largest reduction in oil-water interfacial surface tension per unit concentration (i.e., $slope_i$), and (3) achieves the largest reduction in interfacial tension (i.e., Δwt). Consequently, increased dispersant effectiveness and/or efficiency is reflected by (1) low values for the CMC, (2) high values for the $slope_i$), and (3) high values for the Δwt.

Flume Tests

A major advantage of flume tests is that they can more closely approximate conditions encountered in real-world situations such as wave patterns similar to those at sea can be generated in the test tank and spray applications for dispersant onto a slick can be generated that

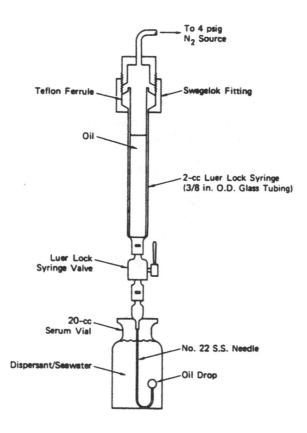

Figure 21. Drop-weight test apparatus. *Source*: Rewick et al., 1981. (Copyright American Petroleum Institute, 1981. Reprinted with permission.)

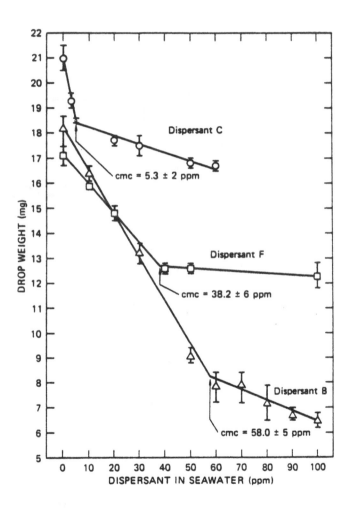

Figure 22. Typical CMC curves for light Arabian crude versus dispersant in seawater with the drop-weight method. Temperature=28°C and salinity=38 ppt. *Source*: Rewick et al., 1984. (Copyright American Society for Testing and Materials, 1984. Reprinted with permission.)

would be similar to those performed from a ship. Two flume tests are discussed below.

Cascading Weir Test (Mackay)--

Mackay et al. (1984) described a test system consisting of a glass-sided flume 8 m long and 27 cm wide with a water depth of 7 cm (Figure 23). Water in the flume passes over weirs with slopes of 20:1 to 100:1 (horizontal:vertical) to simulate wave action. Oil is discharged at a continuous rate onto the water surface, producing a slick that flows down the flume. Dispersant is sprayed onto the oil at prescribed rates (e.g., 1 mL/second). Actions of dispersed oil passing through the flume are visually observed and photographed. Water samples are collected at the end of the flume and analyzed for their oil content to determine dispersant performance as well as oil droplet size. Dispersant-to-oil ratios (DOR) and flow rates can be experimentally altered in the system. The actual dispersion process can be viewed at various points along the length of the flume, which will correspond to various mixing times after dispersant addition. The flume system is also useful for observing vertical rising of oil droplets in a turbulent water column and reformation of a slick.

Delft Hydraulics Flume Test--

Delvigne (1985) presents results of dispersant trials in an even more sophisticated flume system constructed at the Delft Hydraulics Laboratory in the Netherlands (Figure 24). The flume is 15 m long and 0.5 m wide with a maximum water depth of 0.6 m. The walls and bottom consist mainly of glass to allow visual observation as well as direct detection of dispersion parameters by laser beams. The flume is equipped with a programmable wave board for generating nonbreaking and breaking waves, equipment for the withdrawal of water samples at the test section, facilities for the application of oil slicks, and a dispersant spraying system. Additional capabilities and features of the flume include the following: (1) a laser Doppler velocity meter to measure currents, orbital movements, and turbulent velocities in the water mass, (2) a particle sizer for recording the droplet size distribution of the dispersant spray and the dispersed oil droplets, (3) equipment to measure the dispersed oil concentration by either infrared (IR) spectroscopy on samples removed from the flume or absorption values obtained from a flume-crossing laser beam, and (4) determination of the composition of surface oil and dispersed oil and the DOR from gas chromatograph (GC) profiles. While the system is complicated and clearly cannot be installed in areas with space or equipment limitations, it offers distinct advantages for studying dispersion of oil droplets under simulated real-world conditions that include breaking and nonbreaking waves in carefully controlled and reproducible conditions. Consequently, experiments with such a flume system can provide considerable information to evaluate effects of a number of variables and conditions such as generation of nonbreaking and breaking waves, currents, temperature, salinity, oil weathering, slick thickness, dispersant spray droplet size, and droplet impact of dispersant on the surface of the oil slick. It is also possible to measure sizes of dispersed oil droplets in the water column.

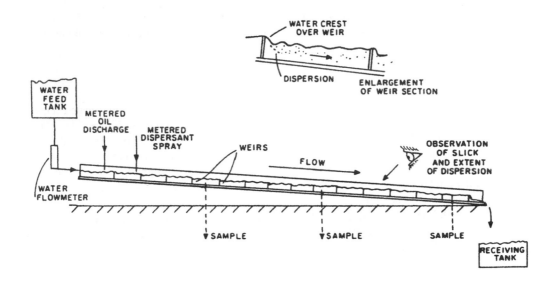

Figure 23. Cascading weir test apparatus. *Source*: Mackay et al., 1984. (Copyright American Society for Testing and Materials, 1984. Reprinted with permission.)

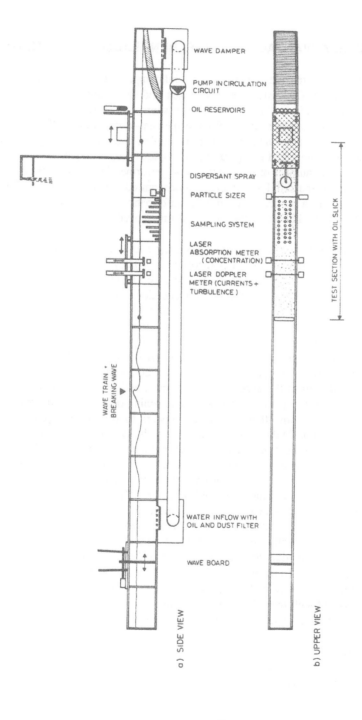

WAVE DAMPER

PUMP IN CIRCULATION CIRCUIT

OIL RESERVOIRS

DISPERSANT SPRAY

PARTICLE SIZER

SAMPLING SYSTEM

LASER ABSORPTION METER (CONCENTRATION)

LASER DOPPLER METER (CURRENTS + TURBULENCE)

WAVE TRAIN + BREAKING WAVE

WATER INFLOW WITH OIL AND DUST FILTER

WAVE BOARD

a) SIDE VIEW

TEST SECTION WITH OIL SLICK

b) UPPER VIEW

Figure 24. Delft Hydraulics flume test apparatus. *Source:* Delvigne, 1985. (Copyright American Petroleum Institute, 1985. Reprinted with permission.)

ADVANTAGES AND DISADVANTAGES OF DIFFERENT LABORATORY TESTS

Table 4 summarizes features and essential procedural components in the testing methods described above. To assist in discussions of advantages and disadvantages of the various procedures, the table includes information for the energy source, relative rating for turbulent energy level, water volume, oil-to-water ratio (OWR), dispersant-to-oil ratio (DOR), dispersant application method, settling-time preceding collection of samples, and relative rating for overall complexity of the testing procedure and its operation. Complexity is a qualitative variable that will include factors such as the number of tests that can be performed in a given period of time with an apparatus, the training and skill level of an operator required to perform a particular test, and the overall cost to acquire a particular apparatus. Information in the table is relevant to discussions of advantages and limitations of the various testing procedures.

Byford and Green (1984) assess the relative advantages and disadvantages of the MNS and Labofina tests using a variety of oils and dispersants. They report reasonably good agreement between test results with the two methods (e.g., see Figure 8). However, the onset of wave dampening in the MNS test could account for occasional differences between results with the methods. Several approaches are proposed as possibilities for reducing effects of wave dampening in the MNS procedure. For example, tests might be conducted at higher temperatures. The air flow rate might be raised to reduce the effect of wave dampening, although this could promote dispersion at unrealistically high energies. Mixing might be introduced with a stirring device as opposed to an air current, although this would create practical difficulties in the design of the experimental apparatus and further complicate the test procedure. Sampling might be completed before wave dampening occurs (e.g., decreasing the duration of air-induced mixing from 10 to 5 minutes appears to remove the problem; P.S. Daling, personal communication). In the MNS test apparatus, the rate of air flow also is critical for obtaining reliable results. Furthermore, satisfactory generation of waves cannot be obtained unless the test apparatus is precisely level. Deviations as small as 5° from the 90° angle of entry in the air supply are thought to cause significant changes in test results. Byford and Green conclude that reproducibility of test results obtained with the MNS test are relatively good, although all tests in which wave dampening was observed in the MNS procedure were excluded from the comparison. Fingas et al. (1987b) list a number of disadvantages encountered in the Environment Canada laboratory with the MNS test procedure: (1) values for dispersant performance are frequently high, due to factors such as a lack of correction for natural dispersibilities of oil (i.e., separate from chemical dispersion); (2) the relative complexity of the apparatus means that performance of multiple testing events in short periods of time is not possible; (3) results can be difficult to replicate with the apparatus; (4) test results can be more variable in the absence of a settling-period after agitation is stopped; (5) wave dampening can be problematic with different oil and dispersant types; (6) viscous shearing of droplets can occur, which can produce incorrect dispersion performance for medium viscosity oils; (7) the positioning of the sampling port in the apparatus is critical for obtaining reproducible results; (8) the air stream likely creates an unrealistically high energy regime; and (9) the high velocity flow

Table 4. Summary of features of laboratory methods to test dispersant performance

Test ID	Reference	Energy Source	Energy Rating[a]	Water Volume (mL)	OWR[b]	Dispersant Application Method	DOR[c]	Settling Time (min)	Complexity Rating[d]
MNS	Mackay and Szeto, 1981	high velocity air stream	3	6000	1:600	dropwise/premix	variable	none	3
Revised Std. EPA	EPA, 1984	pump	3	130,000	1:1300	dropwise	3:100 to 1:4	none	3
oscillating hoop	Fingas et al., 1989a	oscillating hoop	3	35,000	1:175	dropwise/premix	variable	none	3
IFP-dilution	Desmarquest et al., 1985; others	oscillating hoop	1-2	4000-5000	1:1000, then decrease	dropwise	variable	none	2
flowing cylinder	Fingas et al., 1989b	vertical flow of water	1	1000	1:1200, then decrease	premix	1:25	10	2
Labofina rotating flask	Martinelli, 1984	rotating vessel	3	250	1:50	dropwise	1:25	1	1
swirling flask	Fingas et al., 1987a	shaker table	1-2	120	1:1200	premix/dropwise	1:10 to 1:25	10	1
EXDET	Becker et al., 1991	wrist-action shaker	1-2	250	variable	premix/dropwise	variable	none	1
drop-weight	Rewick et al., 1981, 1984	none	0	(NA)[e]	(NA)	water-oil interaction	(NA)	(NA)	2
cascading weir	Mackay et al., 1984	water passing over weirs	2-3	150,000	variable	spray	variable	none	4
Delft flume	Delvigne, 1985	wave paddle	2-4	4,500,000	variable	spray	variable	none	4

[a] Energy Rating: 0 = none; 4 = highest
[b] OWR = oil-to-water ratio (v:v)
[c] DOR = dispersant-to-oil ratio (v:v)
[d] Complexity Rating: 1 = lowest; 4 = highest
[e] (NA) = not applicable

of air can induce high evaporative losses of volatile components in test oils. In addition to the above-mentioned difficulties, an additional disadvantage of the MNS test is the relative complexity of the apparatus and its expense to acquire. Considerable time also is required to not only clean the apparatus but also stabilize water temperature for conduct of a test. The major advantage of the MNS test is that it provides energy in the form of a tangential air stream, which is more representative of real-world wind and water conditions. The air stream also spreads the oil into a thinner film and assists the mixing process between oil and dispersant in a manner similar to that occurring at sea during aerial application of dispersants.

Byford and Green (1984) note that a major advantage of the Labofina test is its speed and simplicity, which allow for as many as 16 test runs to be performed during a normal working day. Byford and Green conclude that rotation speeds in the range of 31-37 RPM in the Labofina test have little effect on results. The parameter that causes the greatest effect is the shape of the conical separatory funnel. Long, narrow flasks produce higher dispersion values than relatively shorter and wider vessels. Also, the size of the orifice in the stopcock is important because it affects the time required to collect a 50-mL sample from the flask. As noted in Figure 19, flask geometry has been standardized for the Labofina procedure in the United Kingdom to minimize flask-related effects on test results. As for additional disadvantages, Byford and Green note that test results may be biased to give low dispersion estimates because oil occasionally can adhere to walls of test flasks. Furthermore, energy input to the test solution during rotation of the flask is quite high (Gillot et al., 1986/87). The high oil-to-water ratio (1:50) also may yield unrealistically high concentrations of both dispersed oil as well as dispersant in the water.

With the Revised Standard EPA test, Payne et al. (1985) observe that results tend to be very operator dependent (i.e., consistent technique used by an operator is critical for minimizing variability in test results). Extreme care is required to apply the dispersant uniformly to the oil surface at a constant and repeatable rate, and the height of the graduated cylinder above the oil surface during introduction of the dispersant is critical for obtaining reasonably consistent values. The mixing energy, which is applied in the form of a water spray system, can be monitored with a pressure meter. However, the exact height of the spray nozzle above the water surface must be carefully controlled. Caution is also required to avoid splashing of oil onto the tank walls. The larger volume of seawater used in the EPA test as opposed to the Labofina and MNS procedures does provide greater dilution rates for dispersed oil droplets and reduces potential scaling problems inherent to the MNS method. However, relatively large volumes of oil-contaminated water are generated, which can result in waste disposal problems. The apparatus is large, not readily portable, and is difficult to clean. Consequently, only a relatively small number of tests can be performed in a normal working day with the Revised Standard EPA apparatus. The test also requires a specially constructed tank and spray system, which means that the system is relatively expensive.

The IFP-Dilution test has a number of advantages including (1) the flow-through design of the apparatus that allows for continuous replacement of water beneath the slick and simulates dilution losses of dispersed oil droplets that occur in field situations, (2) the intended maintenance of the oil slick toward the center of the test solution that minimizes wall-effects on test results, and (3) some variation in energy levels for test solutions can be achieved by varying the rate and stroke-length of the oscillating hoop. Disadvantages include (1) the rather complicated construction design (and, hence, cost) for the apparatus, (2) cleanup of the apparatus at the conclusion of a given experiment is not trivial, and (3) a relatively limited number of experiments can be performed in a normal working day (e.g., 4-5 tests per apparatus for 1-hour runs described by Daling et al., 1990b, and CEDRE, undated; 3 tests for 2-hour runs according to Desmarquest et al., 1985, Bardot et al., 1984, and Bocard et al., 1984).

The flowing cylinder test used at Environment Canada has advantages that include (1) the presence of very low energy regimes in the test apparatus and (2) the flow-through design of the apparatus that allows for continuous replacement of oil-free water beneath the slick and assurance that large, unstable oil drops will return to the surface slick. Disadvantages include (1) the rather sophisticated nature of parts of the testing apparatus (e.g., the peristaltic pump and filtration system), (2) cleanup of the apparatus between experiments is not trivial, and (3) a relatively small number of experiments can be performed in a normal working day.

The Swirling Flask test has advantages including (1) test components are constructed from readily available and inexpensive items, (2) multiple tests can be rapidly and easily performed, and (3) some variation in energy levels for test solutions can be achieved by varying the rate or stroke-length of action on the shaker table. Initial published versions of the test utilized only premixed oil-dispersant solutions, which can be appropriate for evaluating dispersant performance independent of problems associated with the application of dispersant to an oil slick. Recent variations of the test include one- and two-drop dispersant application procedures onto the test slicks (e.g., Fingas et al., 1990), which can provide some information on herding effects of dispersants on overall dispersant performance.

The Drop-Weight test utilizing changes in oil-water interfacial surface tension to estimate dispersant effectiveness has a number of advantages: (1) measurement is obtained for the Critical-Micelle-Concentration (CMC), which is directly related to the dispersion effectiveness that can be imparted by the dispersant agent to the oil, (2) variable mixing energies are avoided in the testing apparatus, (3) wall-effects on test results are avoided, (4) wave dampening is not applicable, and (5) variability associated with different sampling methodologies is avoided. Disadvantages of the Drop-Weight test include (1) the actual amount of dispersed oil is difficult to estimate, (2) effects of numerous environmental variables (e.g., energy levels) on dispersion performance cannot be estimated, (3) the time required for droplet development may be operator dependent and variable, (4) experimental conditions are far removed from real-world spill scenarios, and (5) inherent deviations from a dispersant manufacturer's recommended application method are required to perform the test.

Delvigne (1985) lists the following advantages that are present in tests with flume systems that simulate dispersion processes in real-world situations:

- A more adequate modeling of surface energy due to breaking and nonbreaking waves and, possibly, water currents and wind is obtained.

- Better modeling of energy in the water for dispersion of oil droplets occurs.

- A measurement and control system for energy, such as a laser Doppler velocity meter, can be present.

- The capacity to vary the thickness of the oil slick as well as temperature and salinity are possible.

- Adequate equipment for applying dispersant in concentrated or diluted form, including space systems capable of producing variable droplet impact and variable droplet size, can be present.

- The capacity to evaluate a variety of oil types including weathered or emulsified crude is possible.

- Equipment for measuring oil concentrations in the water column can be included.

- Equipment for determining oil composition can be utilized with experimental test samples.

- Equipment for evaluating rheological properties of the oil (e.g., viscosity, density, and interfacial surface tension) can be utilized.

Ideally, laboratory test systems should be large enough to reduce wall-effects and permit a greater internal range in energy structure and a small viscosity length scale, which would correspond more closely to oceanic circumstances. Such conditions can be achieved in larger test flumes. A further advantage to such flumes is that they can be used to evaluate dispersant performance in the presence of sea ice. None of the other smaller-scale laboratory test systems are easily adaptable for such desired purposes, although limited ice studies have been reported in Byford et al. (1983) and Mackay et al. (1980). Disadvantages to flume systems include (1) size, (2) complexity, (3) cost, (4) large space requirements, (5) generation of very large quantities of oily waste, and (6) very long turn-around and cleanup times between experiments. In summary, major directions for studies in large flume systems will appropriately remain focused toward specific research topics relating to the process of oil dispersion (i.e., natural as well as chemically-mediated) and to evaluate the relevance of smaller-scale laboratory studies and results to larger-scale real-world scenarios.

RAPID FIELD TESTS FOR ESTIMATING DISPERSANT PERFORMANCE

As noted previously, oil that is spilled on water will undergo rapid changes in chemical and physical properties that will affect the ability of oil to be dispersed by chemical agents. Consequently, onsite tests of dispersant performance during spills can be very desirable because information is provided on a real-time basis about the specific oil at the site. Toward this end, objectives for rapid field tests that could be performed onsite include the following:

- the materials for performing a test should be simple, inexpensive, and readily obtainable,

- the test should be simple so that it can be performed in the field by personnel with a minimum of prior training, and

- test results obtained by personnel in the field should be not only quickly available but also reproducible with an acceptable degree of confidence.

The laboratory test methods discussed above do not readily lend themselves to onsite applications in the field. Consequently, relatively simple field tests have been developed that have the potential to provide rapid, qualitative information regarding an oil's dispersibility. In contrast to the tests described above, however, these rapid field tests are inherently limited in the scope of information that they can provide because of their necessary simplicity for use in the field.

Five rapid, relatively easy tests that have the capacity to be utilized in field situations are discussed in the following sections. Each section also includes a brief discussion of limitations inherent to the test, which dictate that results are most appropriately considered in qualitative rather than quantitative terms.

EPA Field Dispersant Effectiveness Test

The EPA Field Dispersant Effectiveness Test is described in Diaz (1987). Essential apparatus for the test include a standard 0.5-inch test tube, ruler, stopwatch, and flashlight. Seawater (synthetic) is added to the test tube to a height of 5 cm. Ten drops of oil followed by one drop of dispersant (i.e., an expected dispersant-to-oil ratio or DOR of 1:10) are added to the top of the seawater in the test tube. The tube is stoppered and shaken for 1 minute in a vertical orientation at 120 cycles/minute with a 4-inch stroke. The tube is then maintained in a stationary, vertical position for 10 minutes. After this settling-time, the tube is set on top of a shielded flashlight in which the light is passed through an opaque shield with a 0.5-inch opening and an O-ring is moved along the length of the tube until light is no longer visible from the side through the seawater-oil mixture. The measured height of this final O-ring placement (L) is recorded, and the percent dispersion (D) is estimated as:

$$D \text{ (in \%)} = [(5-L)/5] \times 100 \tag{6}$$

where 5 (in cm) is the initial height of the seawater in the tube and L (in cm) is the height from the bottom of the tube to the final O-ring position. While quite simple, the test is subject to the following limitations: (1) the additions of oil and dispersant as droplets as opposed to actual volume measurements can yield varying DOR values depending on the number of drops per unit volume for the oil and dispersant, (2) the intensity of the illumination from the flashlight can affect the transmission of light through the seawater-oil mixture, and (3) the colors of different oils can influence light transmission through the seawater-oil mixture. Items (2) and (3) can affect final placement of the O-ring along the length of the test tube and, hence, the estimated value for dispersion performance.

API Field Dispersant Effectiveness Test

The API Field Dispersant Effectiveness Test (S.L. Ross Environmental Research Limited, 1989) is intended to mimic the Warren Spring Laboratory rotating flask test. The main apparatus for the test is a modified battery or antifreeze tester, which consists of a glass tube that is 2 cm in diameter and 21 cm long. A solid rubber stopper is inserted into the open end of the tube while the rubber drain hose on the other end is clamped and left in place. A 50-mL volume of seawater is added to the tube, followed by 1 mL of a premixed oil-dispersant mixture (DOR=1:20). The tube is then rotated end-over-end in the vertical plane at 30 RPM for 2 minutes by hand. After a 5-minute settling-time with the tube in a vertical position and the drain hose at the bottom, a 25-mL sample is removed through the drain hose. The sample is extracted in a flask or narrow-neck bottle by shaking for 30 seconds with 25 mL of toluene. After a 30-minute period to allow for phase separation in the extraction flask/bottle, an additional 225 mL of toluene is added to the flask/bottle. A sample of the final toluene is then collected. Performing the 1:10 dilution of the toluene in the flask/bottle after the shake-extraction step minimizes turbidity in the toluene phase. Oil standards in toluene are prepared from the oil used in the test, with standard concentrations made at 2000, 1500, 1000, and 500 ppm. The 2000-ppm standard is prepared by dissolving 1 mL of the oil in 500 mL of toluene. The 1500-, 1000-, and 500-ppm standards are prepared as dilutions of the 2000-ppm standard. To categorize the dispersion performance for an oil in the test, the color of the toluene extract from the dispersed sample is visually compared to that of the oil standards. While relatively simple, the test is limited by the fact that estimation of dispersant performance is based on visual comparison of a sample extract to a series of oil standards, although the reference citation for the method does show good agreement between results with the API test and the Warren Spring Laboratory rotating flask test for two oils and three dispersants. Preparation of oil standards at additional concentrations also might improve estimation of dispersant performance, although increasing the number of standards will increase not only the time required but also the overall number of items needed to perform the test in field situations.

Mackay Simple Field Test

The Mackay Simple Field Test was developed by D. Mackay (Abbott, 1983). For the test, 1000 mL of seawater (or a 30 g/L solution of NaCl) are added to a 1-liter volumetric flask. A 10-mL volume of oil is added to the surface of the water, followed by 1 mL dispersant. The volumetric flask is allowed to stand for 1 minute to permit the dispersant to soak into the oil. The flask is then rotated and maintained at a 140° angle (i.e., the neck of the flask is at a 5-o'clock position until the oil and air float out of the neck of the flask. The flask is then returned to a vertical position. This rotation sequence is repeated 30 times. Following all rotations, the flask is maintained in an upright position. Visual measurements of the depth of the oil layer returning to the water's surface in the neck of the flask are taken at specified time intervals for

15 minutes. Comparison of the heights of the oil-water interface in the neck before and after the dispersion test is used to estimate the volume of oil dispersed in the test. The volume fraction of oil remaining in the water after 3 minutes is used to estimate the performance of the dispersant agent. The test is very easy to perform and requires a very simple, readily available apparatus. However, adherence of oil to the walls of the volumetric flask and the ability to accurately measure the thickness of the oil layer in the neck of the flask are limitations that can affect final estimates of dispersion performance.

Pelletier Screen Test

A rapid, comparative test based on visual determination of dispersant performance has been developed by Pelletier (1987). In the test, 20 mL of synthetic seawater and 0.1 mL of oil are added to a 25-mL vial. A 1-cm deep vortex is created in the seawater-oil mixture by a magnetic stirrer, and 0.05 mL of a dispersant is added to the center of the vortex. The stirring rate is increased to 2000 RPM for 60 seconds, followed by a settling-time of 60 seconds. The extent of dispersion of the oil is visually estimated and classified from A (complete) to E (no dispersion). Results of the test are obviously very qualitative and dependent on the subjective determination of the operator.

Fina Spill Test Kit

Fina (undated) markets a spill test kit for dispersant performance. The test adds 2 mL of oil to 100 mL of synthetic seawater in a graduated cylinder. Dispersant amounts of 0.1 and 0.2 mL (i.e., dispersant-to-oil ratios of 1:20 and 1:10, respectively) are added to separate test cylinders. The cylinders are shaken for 10 seconds and then allowed to stand for 30 seconds. Dispersant performance is estimated by comparing the color of the dispersion to a color scale supplied with the kit. Serious limitations can be encountered in this test because different colors will occur for different oils depending on not only the type of oil tested but also the weathering state of the oil. Furthermore, the amount of ambient light also may affect the color of the oil dispersion.

SECTION 5

FIELD TESTS OF DISPERSANT APPLICATIONS

The major purposes of this book are to update information on the mechanism of action of chemical dispersants and factors influencing performance as well as evaluate laboratory procedures for testing performance. A detailed evaluation of studies related to performance of dispersants in field trials is not a major objective of this book. However, brief consideration of results from field studies will be presented because of the relevance of this topic toward the conduct and application of laboratory test results for dispersant use in real-world situations.

A review of field studies of dispersant applications is presented in Fingas (1989). Table 5 summarizes information for 106 separate offshore experimental spills. Estimates for short-term dispersant performance are available for 25 of the spills in the table, with performance values ranging from 0 to 100%. In addition to onsite visual observations at spill sites, techniques for estimating dispersant performance in field tests have included measurements of oil in subsurface water samples, measurements of oil remaining on the water's surface after application of a dispersant, dispersant amounts or distributions, and remote sensing to visually observe results and/or quantify areas of surface oil. Most estimates for dispersant performance in field studies have been developed using data from subsurface concentrations of oil in water columns.

Results describing successful application of chemical dispersants to a series of experimental oil spills in carefully designed experiments offshore of Long Beach, California are presented in McAuliffe et al. (1981). A total of nine spill events with Prudhoe Bay crude oil (10 or 20 barrels of oil/event) were performed in a 2-day period in September 1979. Seven spills were treated with chemical dispersants: three were sprayed with an unidentified dispersant from a fixed-wing aircraft (DC-4), three with the same dispersant from a boat, and one with a second unidentified dispersant from the boat. Two spills were not treated with dispersants and served as controls. Documentation of chemical dispersion of oil in the test spills was implemented by aerial observations and photography, underwater observations and photography, and chemical analyses of more than 900 water samples from the upper 9 meters of impacted water columns. Observations and collected water samples were obtained within 6 hours of the release of the oil or treatment of the slick with the dispersant agent. Both aerial and underwater observations established that dispersion of oil into the upper water column did occur in treated spill events. Chemically-dispersed oil appeared as yellow-brown or yellow-green suspensions in the near-surface waters in association with slicks. Little or no yellow coloration in the water was observed beneath slicks that were not well dispersed. Chemical analyses of water samples detected oil concentrations as high as 55 ppm in the upper 2 meters of water columns beneath chemically-dispersed slicks. Comparable measurements for oil beneath untreated slicks always yielded concentrations that were less than 1 ppm or undetected (i.e., less than or equal to 0.06 ppm). McAuliffe et al. noted that thicknesses of oil slicks on the water's surface were not uniform, and that the degree of chemical dispersion of oil was much greater in thicker portions

Table 5. Summary of at-sea trials for dispersant performance[a]. *Source*: Fingas, 1989. (Copyright American Society for Testing and Materials, 1989. Reprinted with permission.)

Location/ Identifier	Reference	Year	Number	Oil Type	Spill Amount, m^3	Dispersant	Application Method	Dose Rate, D:0	Sea State	Claimed Effectiveness, %
North Sea	Cormack and Nichols [1,2]	1976	1	Ekofisk	0.5	10% conc.	ship, WSL	...	1	...
			2	Kuwait	...	10% conc.	ship, WSL	1:20	2–3	100
Wallops Island	McAuliffe et al. [1,3]	1978	3	Murban	1.7	Corexit 9527	helicopter	1:5	1	...
			4	La Rosa	1.7	Corexit 9527	helicopter	1:5	1	...
			5	Murban	1.7	Corexit 9527	helicopter	1:11	1	100
			6	La Rosa	1.7	Corexit 9527	helicopter	1:11	1	50
South California	Smith et al. [4]	1978	7	North Slope	1.7	Control later Corexit 9527	control then helicopter	>1:5	0–1	...
			8	North Slope	3.2	Corexit 9527	airplane, Cessna	>1:5	0–1	...
			9	North Slope	1.7	Recovery + Corexit 9527	helicopter	>1:5	0–1	...
			10	North Slope	0.8	BP1100WD	ship, WSL	>1:5	0–1	...
			11	North Slope	0.8	Corexit 9527	ship	>1:5	0–1	...
South California	Smith et al. [4]	1978	12	North Slope	3.2	Corexit 9527	airplane, Cessna	>1:5	1–2	...
			13	North Slope	0.8	Corexit 9527	ship	>1:5	1–2	...
			14	North Slope	0.8	BP1100WD	ship, WSL	>1:5	1–2	...
			15	North Slope	0.6	several, demonstration	several, demonstration	...	1–2	...
Victoria	Green et al. [1,6]	1978	16	North Slope	0.2	10%, 9527	ship, WSL	1:1	2	...
			17	North Slope	0.4	10%, 9527	ship, WSL	1:1	1	...
			18	North Slope	0.2	10%, 9527	ship, WSL	1:1	1	...
Long Beach	McAuliffe et al. [1,5]	1979	19	Prudhoe Bay	1.6	control	control	...	2–3	0.5
			20	Prudhoe Bay	1.6	2% conc.	ship	1:67	2–3	8
			21	Prudhoe Bay	1.6	2% conc.	ship	1:67	2–3	5
			22	Prudhoe Bay	3.2	conc.	airplane, DC-4	1:20	2–3	78
Long Beach	McAuliffe et al. [1,5]	1979	23	Prudhoe Bay	1.6	conc.	airplane, DC-4	1:25	2–3	45
			24	Prudhoe Bay	1.6	control	control	...	2–3	1
			25	Prudhoe Bay	3.2	conc.	airplane, DC-4	1:27	2–3	60
			26	Prudhoe Bay	1.6	2%	ship	1:11	2–3	11
			27	Prudhoe Bay	1.6	2%	ship	1:11	2–3	62
Mediterranean, Protecmar I	Bocard et al. [7]	1979	28–41	light fuel	3 each	BP1100X BP1100WD Finasol OSR-5 Corexit 9527	ship, helicopter, various and airplane CL215	...	1–3	...
Mediterranean, Protecmar II		1980	42–49	light fuel	1–5.5	BP1100X BP1100WD Finasol OSR-5 Corexit 9527	ship, helicopter, various and airplane CL215	...	1–3	...
Mediterranean Protecmar III	Bocard and Gatellier [1,7,8]	1981	50	light fuel	6.5	Dispolene 325	airplane, CL215	1:3	1–2	50
			51	light fuel	6.5	Shell	airplane, CL215	1:3	2–3	...
			52	light fuel	6.5	control	control	...	1–2	...
Newfoundland	Gill et al. [9]	1981	53	ASMB	2.5	control	control	...	1	...
			54	ASMB	2.5	Corexit 9527	airplane, DC-6	1:10	1	...

83

Table 5 (continued)

Location/ Identifier	Reference	Year	Number	Oil Type	Spill Amount, m^3	Dispersant	Application Method	Dose Rate, D:0	Sea State	Claimed Effectiveness, %
Norway	Lichtenthaler and Daling [1,10]	1982	55	Statfjord	0.2	control	control	...	2–3	0.6
			56	Statfjord	0.2	10% conc.	ship	1:10	2–3	6
			57	Statfjord	0.2	10% conc.	ship	1:10	2–3	17
			58	Statfjord	0.2	control	control	...	2–3	2.6
			59	Statfjord	0.2	10% conc.	ship	1:17	2–3	19
			60	Statfjord	0.2	10% conc.	ship	1:18	2–3	22
			61	Statfjord	0.2	10% conc.	ship	1:13	2–3	2
North Sea	Cormack [1,11]	1982	62	Arabian	20	control	control	...	1	...
			63	Arabian	20	Corexit 9527	airplane, Islander	1:2	1	...
			64	Arabian	20	Corexit 9527	airplane, Islander	1:4	1	...
Mediterranean Protecmar V	Bocard et al. [1,12]	1982	65	light fuel	3	10% Dispolene 325	ship	1:2	3	...
			66	light fuel	5	Dispolene 325	airplane, CL215	1:2.4	3	...
Protecmar V	Bocard et al. [1,12]	1982	67	light fuel	5	Dispolene 325	ship	1:2.8	2	...
			68	light fuel	5	Dispolene 325	airplane, CL215	1:2.8	2	...
			69	light fuel	3.5	Dispolene 325	ship	1:2.6	1–2	...
			70	light fuel	4	Dispolene 325	helicopter	1:2.9	1–2	...
			71	light fuel	2	premixed	premixed	1:20	1–2	40–50
			72	light fuel	5	control	control	...	2	...
Holland	Delvigne [1,13]	1983	73	Statfjord	2	control	control	...	1–2	2
			74	light fuel	2	control	control	...	1–2	2
			75	Statfjord	2	control	control	...	1	2
			76	Statfjord	2	Finasol OSR-5	airplane	1:10–30	1	2
			77	light fuel	2	Finasol OSR-5	airplane	1:10–30	1	2
			78	Statfjord	2	Finasol OSR-5	premixed	1:20	2–3	100
Holland	Delvigne [1,13]	1983	79	light fuel	2	control	control	...	2–3	2
			80	Statfjord	2	Finasol OSR-5	airplane	1:10–30	1–2	2
			81	Statfjord	2	Finasol OSR-5	airplane	1:10–30	1–2	2
Halifax	Swiss and Gill [1,14,15]	1983	82	ASMB	2.5	Corexit 9527	helicopter	1:20	1	2.5
			83	ASMB	2.5	control	control	...	1	1
			84	ASMB	2.5	Corexit 9550	helicopter	1:10	1	13
			85	ASMB	2.5	control	control	...	1	1
			86	ASMB	2.5	BP MA700	helicopter	1:10	2–3	10–41
			87	ASMB	2.5	control	control	...	2–3	7
Norway	Lichtenthaler and Daling [16]	1984	88 '	Statfjord	10	control	control	...	1	...
			89	Statfjord	10	Corexit 9527	airplane, Islander	1:75	1	...
			90	Statfjord	10	control	control	...	2	...
Norway	Lichtenthaler and Daling [16]	1984	91	Statfjord	10	Corexit 9527	airplane	1:80	2	...
			92	Statfjord	12	Corexit 9527	premixed	1:33	2	...
			93	Statfjord	10	Corexit 9527	airplane	1:50	...	...
Brest, Protecmar VI	Bocard [7,17]	1985	94	fuel oil	5	control	control	...	1	...
			95	fuel oil	28	Dispolene 355	helicopter	1:9	1	...
			96	fuel oil	part of above	Dispolene 355	ship-spray	1:9	1	...
			97	fuel oil	part of above	Dispolene 355	ship-aerosol	1:9	1	...
Haltenbanken	Sørstrøm [19]	1985	98	topped Statfjord crude	12.5	control	...	...	1–2	...
			99	topped Statfjord crude	12.5	Finasol	premixed, injected 3 m below surface	1:50	1–2	...

Table 5 (continued)

Location/ Identifier	Reference	Year	Number	Oil Type	Spill Amount, m³	Dispersant	Application Method	Dose Rate, D:0	Sea State	Claimed Effectiveness, %
Haltenbanken	Sørstrøm [19]	1985	100	topped Statfjord crude	12.5	control	...	...	1–2	...
			101	topped Statfjord crude	12.5	alcopol (demulsifier)	premixed	250 ppm	1–2	...
Beaufort Sea	Swiss et al. [20]	1986	101 (CA)	topped Federated crude	2.5	control	...	...	1–2	...
			102 (CB)	topped Federated crude	2.5	control	...	...	1–2	...
			103 (C1)	topped Federated crude	2.5	BP MA700	helicopter	1:10	2–3	...
			104 (T1)	topped Federated crude	2.5	BP MA700	helicopter	1:1	2–3	...
			105 (T2)	topped Federated crude	2.5	Corexit CRX-8	helicopter	1:1	2–3	...
			106 (C)	topped Federated crude	2.5	control	...	...	2–3	...

[a] Abbreviations: ASMB—Alberta Sweet Mixed Blend, conc.—concentrate, and WSL—Warren Spring Laboratory.

of treated slicks.

Lichtenthaler and Daling (1985) present results from field studies conducted to evaluate effects of a chemical dispersant on oil slicks released off the Norwegian coast. Six spill events (10 m^3 oil/event) were generated with a simulated topped Statfjord crude oil. Two spills were controls (i.e., no dispersant application), one spill received an aerial dispersant application one hour after the oil release, two spills received dispersant application two hours after oil release, and one spill involved a premixed blend of dispersant and oil. Corexit 9527 was the dispersant used in all tests. Aerial surveillance and photography as well as chemical analyses of water-column samples (0.5, 1, 2, and 3 meter depths) and surface oil samples were conducted for periods up to 12 hours after a spill event to evaluate dispersion performance for the various slicks. Results were similar to those described above for the Southern California field study. For example, aerial observations/photography indicated that chemically-dispersed oil appeared as plumes with yellow-brown to green-white colors in the near-surface waters in association with the slicks. Chemical analyses of water samples detected oil concentrations in the upper 2 meters of water columns that could be as high as 60 ppm for the slicks treated by aerial dispersant application and 185 ppm for the test involving the premixed oil-dispersant addition to the water's surface. Water concentrations of oil beneath the untreated slicks were mainly in the ppb range, except for a small number of samples at 0.5-1 m depths that contained unstable dispersed oil droplets due to breaking waves. While the preceding observations and measurements were for events relatively soon after the spill events or treatments (i.e., within 12 hours), the authors also note that secondary, longer-term dispersion of oil (e.g., up to 1-2 days after a spill and treatment) was observed.

The preceding results provide examples of successful dispersion of oil in slicks treated in the field. However, many of the studies summarized in Table 5 report poor dispersion of oil in the short-term following treatment efforts. Poor dispersion in field situations can result from a variety of factors, such as ineffective application of a dispersant to a slick (see section on dispersant application). Mackay and Chau (1986/87) also note that dispersant performance on oil at sea (as well as in the laboratory) is influenced by the chemical natures of the dispersant and oil, their relative amounts, the extent of evaporation and water-in-oil emulsification (i.e., mousse formation) for the oil, differential spreading of the oil into varying thicknesses, and the microscopic mixing processes that occur as dispersant lands on and penetrates into oil. Interpretation of dispersion results from measurements of oil in water samples also can be complicated by irregular patterns of dispersed-oil plumes beneath slicks. Brown et al. (1987) note that extensive sampling of such plumes is necessary to accurately account for the spatial heterogeneity of dispersed oil beneath a slick and obtain true estimates of dispersant performance. Additional complications for estimating dispersant performance can be caused by herding of surface oil by dispersants (e.g., Merlin et al., 1989; also see discussion in section on application of dispersants). Herding generally occurs soon after application of dispersant to an oil slick and has occasionally been misinterpreted as evidence supporting chemical dispersion. Resurfacing of dispersed oil following a dispersant application also can lead to over-estimation of dispersion unless resurfacing is monitored and corrected for in field situations. In summary, numerous precautions must be taken into account in evaluating information from field tests for estimating performance of chemical dispersants in field tests.

From available information for field trials, at-sea tests, and spills-of-opportunity, it is clear that varying conclusions can occur regarding the performance of chemical dispersants in real-world spill events. Information on dispersant performance in such situations is often problematic because (1) dispersants are often used on minor spills in which there is not time to adequately monitor the spill, (2) obtaining the necessary surface and subsurface measurements for oil in the water column is complicated and difficult, (3) mass-balance estimates for the oil are difficult to achieve, and (4) most of the data for dispersant-to-oil ratios in field situations represents estimates for the total volumes of oil spilled and dispersant applied as opposed to the actual amounts involved in dispersant-oil interactions in the particular spill situation. Furthermore, it is generally not possible to determine how much dispersant is successfully applied to the slick, at what concentration, and in what droplet size range. However, some spills have been successfully treated with dispersants, while others have not. From results of field trials, it appears that processes responsible for mixing of a dispersant into an oil as well as changes in the physical properties of the oil are critically important for dispersion of the oil.

From results of field trial tests, the concentrations of dispersed oil in the water column often appear to be lower than might be expected. Bocard et al. (1984) discussed dissolution processes in at-sea trials, and suggested that laboratory tests incorporate flow-through designs because closed vessels cannot duplicate dilution processes operating in natural systems. That is, simultaneous measurements of rates of dispersion of oil and its subsequent advective loss from test solutions in the laboratory have definite environmental relevance. This dilution factor also should be considered in toxicity tests in which marine test organisms are subjected to varying concentrations of dispersed oil that reflect dilution processes. The IFP-Dilution and flowing cylinder tests described in the section on laboratory testing methods incorporate time-series dilutions of dispersed oil droplets that attempt to simulate natural dilution processes.

Efforts have also been initiated to utilize remote sensing to assess dispersant treatment of oil spills and monitor the growth and transport of spills (e.g., Goodman and MacNeill, 1984; Goodman and Morrison, 1985; McColl et al., 1987). It is still important, however, to continue on-site chemical measurements and visual observations. Such efforts will include collecting water and surface slick samples for chemical determinations of not only dispersant-to-oil ratios but also quantities of oil dispersed into a water column. Some investigators have recommended utilization of mass balance estimates for surface oil before and after dispersant application to estimate dispersion performance. Measurements of dispersant-to-oil ratios in actual aerial applications are critically needed. In several field tests, concentrations and sizes of dispersant droplets have been estimated by collection of dispersant droplets on paper targets. This is an area that still requires considerable evaluation because dispersant performance in field applications will depend on short-term, dispersant droplet/oil interactions. Field estimates of the diameters of dispersed oil droplets in water columns also need to be completed; these have not been done successfully to date in field operations. Further correlation between field and laboratory studies will clearly require this type of information.

SECTION 6

SUMMARY AND RECOMMENDATIONS - LABORATORY STUDIES

Chemical dispersion of oil into water in laboratory studies involves complex interactions between many variables including the chemical and physical properties of an oil and dispersant agent(s), the method(s) of application (and its effectiveness) and mixing of a dispersant with the oil, the source and magnitude of mixing energy available to the system, the dispersant-to-oil ratio, the oil-to-water ratio, temperature, and the salinity of the aqueous medium. Extrapolation of results from dispersion performance studies in a laboratory to field situations must take into account additional complicating variables including rapid changes that occur in properties of the oil with time (i.e., natural weathering), field application methods and logistics, ambient weather and/or meteorological conditions, and local sea-state or oceanographic conditions (e.g., wave heights, currents, turbulent mixing regimes, etc.). The breadth of these variables make it unlikely that any single laboratory test will ever be completely suitable to quantify performance of chemical dispersant agents for all possible environmental scenarios. Many laboratory test results should more realistically be utilized to apply relative rankings to performance by various dispersant agents, including possible assignment of "acceptable/unacceptable" status to individual dispersants.

While often confusing and/or contradictory, studies conducted with the variety of laboratory testing methods have yielded much information and knowledge about dispersant performance and their mechanisms of action. Adoption of certain experimental protocols (e.g., selection of specific reference oils and dispersants, consideration of the weathered state of the test oil(s), specific oil-to-water ratios, selection or not of designated settling-times to be used in the conduct of experiments, and taking into account natural dispersibilities of oils in a given test) also may lead to closer agreement in performance results between testing methods. However, further advances in testing methodologies remain to be developed and refined, particularly as they relate to the environmental relevance and performance of dispersant agents. For example, there remains room for improvement in approaches used to generate environmentally-relevant mixing energies in laboratory studies. The circulating pump used in the Revised Standard EPA test may cause problems because of the additional interfaces created by shear effects. The spraying method in the test simulates spray boat applications, but is difficult to quantify and leads to poor reproducibility of results. Mixing energy in a separatory funnel (e.g., the Labofina and Warren Spring Laboratory tests) is also difficult to quantify and relate to open sea conditions. The oscillating hoop and MNS tests may be more appropriate for specific environmental conditions, although it has been suggested that energy levels may be too high for at least the MNS apparatus. Flume tests (e.g., the Delft Hydraulics flume and cascading weir designs) provide the highest degree of environmental relevance, but are complex to operate and require long turn-around times between experiments. Flow-through tank designs (i.e., that remove oil droplets) will better simulate the advective removal and dilution of dispersed oil droplets in real-world spill situations. The IFP-Dilution and flowing cylinder tests are attractive in this respect. At the same time, it is desirable that a laboratory testing method be simple,

88

require equipment that is relatively easy to acquire and/or fabricate, require a minimum of operator training and sophistication, and allow for the conduct of a reasonably large number of tests yielding quantifiable results in an acceptably short period of time. The Swirling Flask test and the Labofina/Warren Spring Laboratory test appear most suitable to the latter requirement.

Certain laboratory studies do suggest that continued investigation and analysis of dispersed oil droplet sizes can be useful for explaining differences in energy levels and estimates of dispersion performance in different laboratory testing systems, which could lead to improved, standardized test designs. In general, laboratory experiments also have not been designed to evaluate the effects of herding of oil on dispersion results. Current testing methodologies remain inadequate for investigating dispersion in thin versus thick slicks. Slick thickness and spatial extent are important considerations for dispersant applications at sea because slicks are usually nonuniform in thickness and distribution on the water's surface. It appears that dispersion of oil slicks by chemical dispersants will be most successful with slick thicknesses in the range of 100-1000 um (e.g., slick thicknesses greater than 1000 um are unlikely to occur on large spatial scales and thicknesses less than 100 um will be subject to extreme herding reactions by dispersants; G.P. Lindblom, P.S. Daling, personal communications). The effects of the chemical compositions of oils (including changes due to their weathering), photochemical oxidation, and water-in-oil emulsification (i.e., mousse formation) still have not been fully evaluated. Additional efforts appear to be appropriate in the latter areas to further knowledge of overall dispersion processes (i.e., both natural and chemically-mediated). It also may be appropriate to design dispersant application systems for laboratory test methods that more closely model dispersant droplet sizes and velocities that would be encountered in air-deployed dispersant operations in field situations. At the same time, however, minimizing variables such as application artifacts (e.g., by using premixed oil-dispersant solutions) may be appropriate for laboratory tests designed to evaluate the relative performance of different commercial dispersant products. More work on dispersant applications in ice-filled waters also might be reasonable.

From the standpoint of using chemical dispersants for mitigating effects of oil spills in real-world situations, development and refinement of application techniques and protocols for applying dispersants in the field remain as critical needs. Successful application of chemical dispersants in field situations continues to be problematic. Further studies in areas of application technologies are definitely warranted.

Attempts to model oil-dispersant behavior should be continued. Attempts in this regard have been made (e.g., Mackay, 1985; Daling et al., 1990b and c). Further laboratory (and field) studies should be designed to evaluate effects of important parameters for model considerations. For accurate validation of models, components that need to be more carefully measured in future laboratory and field studies include total volumes of oil, densities of oil, total spill areas, the fractions of slick areas that are thick, the average diameters of dispersant droplets, the overall volumes of dispersant used for applications, the fractions of dispersant that are successfully applied to thicker portions of slicks, dispersant performance values, temperature, and salinity.

A final recommendation for laboratory studies would be to combine results of oil dispersion tests with toxicity information into some common mathematical formulation that could be used to provide better criteria for overall dispersant performance and appropriateness for use in real-world spill situations. One approach used by Anderson et al. (1985) generates a value designated as the relative effective toxicity (RET):

$$RET = (DOR_{90} \times 10^4)/LC_{50} \tag{7}$$

where DOR_{90} is the dispersant-to-oil ratio that gives a dispersion value of 90% (in the MNS test) and LC_{50} is the concentration of dispersant required to cause 50% mortality in test organisms after 96 hours of exposure. RET values could be used to determine use-criteria for dispersant agents based on both dispersion performance and toxicity. In toxicity tests in Anderson et al., 96-h LC_{50} values were between 1 and 10 ppm for seven dispersant products, between 15 and 17 ppm for three products, and between 29 and 32 ppm for two products. Two other dispersants were identified as non-toxic because their 96-h LC_{50} values were in excess of 200 ppm. By comparison, the 96-h LC_{50} value for a standard toxicant, dodecyl sodium sulfate, was 8.7 ppm. A 1:20 (DOR) dispersion of Corexit 9527/Prudhoe Bay crude oil produced a 96-h LC_{50} of 4 ppm, which was approximately one-tenth the 96-h LC_{50} value for Corexit 9527 alone. By combining dispersant performance results and toxicity, it is possible to rank dispersant products and provide information on the dose rate to generate an effective oil dispersion that can be balanced against potential toxicity effects. Thus, a more toxic dispersant might be appropriate to use in a particular spill event, if it could be applied at concentrations significantly lower (10-100X) than a less toxic but less "effective" dispersant. Finally, the relative cost of different dispersant agents also might be factored into such an equation.

SECTION 7

REFERENCES

Abbott, F.S. 1983. A simple field effectiveness test for dispersants. Spill Technology Newsletter, September-October 1983. Environment Canada.

Allen, A.A. 1991. Controlled burning of crude oil on water following the grounding of the EXXON VALDEZ. Proc. 1991 International Oil Spill Conference, 4-7 Mar. 1991, San Diego, CA. American Petroleum Institute, Washington, D.C. pp. 213-216.

Anderson, J.W., D.L. McQuerry and S.L. Klesser. 1985. Laboratory evaluation of chemical dispersants for use on oil spills at sea. Environmental Science and Technology, 19: pp. 454-457.

Anonymous. 1984. Guidelines on the Use and Acceptability of Oil Spill Dispersants. Second Edition, EPS 1-EP-84-1, March, 1984, prepared by the Technical Services Branch, Environmental Protection Programs Directorate, Environmental Protection Service, Environment Canada.

Bardot, C., C. Bocard, G. Castaing, and C. Gatellier. 1984. The importance of a dilution process to evaluate effectiveness and toxicity of chemical dispersants. Proc. Seventh Annual Arctic Marine Oilspill Program Technical Seminar, Environmental Protection Service, Environment Canada. pp. 179-201.

Becker, K.W. and G.P. Lindblom. 1983. Performance evaluation of a new versatile oil spill dispersant. Proc. 1983 Oil Spill Conference, 28 Feb.-3 Mar. 1983, San Antonio, TX. American Petroleum Institute, Washington, D.C. pp. 61-64.

Becker, K.W., L.G. Coker, and M.A. Walsh. 1991. A method for evaluating oil spill dispersants: Exxon Dispersant Effectiveness Test (EXDET). Report presented at Oceans '91 Conference, 1-3 Oct. 1991, Honolulu, HI. 5 p.

Belk, J.L., D.J. Elliott, and L.M. Flaherty. 1989. The comparative effectiveness of dispersants in fresh and low salinity waters. Proc. 1989 Oil Spill Conference, 13-16 Feb. 1989, San Antonio, TX. American Petroleum Institute, Washington, D.C. pp. 333-336.

Belore, R. 1987. A study of dispersant effectiveness using an ultra-uniform drop-size generator. Proc. Tenth Arctic and Marine Oilspill Program Technical Seminar, Conservation and Protection, Environment Canada. pp. 357-371.

Bobra, M. 1990. A study of the formation of water-in-oil emulsions. Proc. Thirteenth Arctic Marine Oil Spill Program Technical Seminar, 6-8 June 1990, Edmonton, Alberta. Environment Canada, Ottawa, Ontario. pp. 87-117.

Bobra, M. and S. Callaghan. 1990. A Catalogue of Crude Oil and Oil Product Properties (1990 Version). Environment Canada Publication No. EE-125, Environment Canada, Ottawa, Ontario, Canada. 542 p.

Bobra, M. 1991. Water-in-oil emulsification: A physicochemical study. Proc. 1991 International Oil Spill Conference, 4-7 Mar. 1991, San Diego, CA. American Petroleum Institute, Washington, D.C. pp. 483-488.

Bocard, C., G. Castaing, and C. Gatellier. 1984. Chemical oil dispersion in trials at sea and in laboratory tests: the key role of dilution processes. In: Oil Spill Chemical Dispersants: Research Experience and Recommendations, ASTM STP 840. (T.E. Allen, ed.). American Society for Testing and Materials, Philadelphia. pp. 125-142.

Bridie, A.L., T.H. Wanders, W. Zegveld, and H.B. van der Heijde. 1980. Formation, prevention and breaking of sea water in crude oil emulsions 'chocolate mousses'. Marine Pollution Bulletin, 11: 343-348.

Brochu, C., E. Pelletier, G. Caron, and J.E. Desnoyers. 1986/87. Dispersion of crude oil in seawater: The role of synthetic surfactants. Oil & Chemical Pollution, 3: 257-279.

Brown, H.M., R.H. Goodman, and G.P. Canevari. 1987. Where has all the oil gone? Dispersed oil detection in a wave basin and at sea. Proc. 1987 Oil Spill Conference. American Petroleum Institute, Washington, D.C. pp. 307-312.

Buist, I.A. and S.L. Ross. 1986/87. A study of chemicals to inhibit emulsification and promote dispersion of oil spills. Oil & Chemical Pollution, 3: 485-503.

Buist, I., S. Potter, D. Mackay, and M. Charles. 1989. Laboratory studies on the behavior and cleanup of waxy crude oil spills. Proc. 1989 Oil Spill Conference, 13-16 Feb. 1989, San Antonio, TX. American Petroleum Institute, Washington, D.C. pp. 105-113.

Byford, D.C. 1982. The development of dispersants for application at low temperature. Proc. Fifth Arctic Marine Oilspill Program Technical Seminar. pp. 239-254.

Byford, D.C., P.J. Green, and A. Lewis. 1983. Factors influencing the performance and selection of low-temperature dispersants. Proc. Sixth Arctic Marine Oilspill Program Technical Seminar, 14-16 June 1983, Edmonton, Alberta, Canada. Environmental Protection Service, Environment Canada. pp. 140-150.

Byford, D.C. and P.G. Green. 1984. A view of the Mackay and Labofina laboratory tests for assessing dispersant effectiveness with regard to performance at sea. In: Oil Spill Chemical Dispersants, Research Experience and Recommendations, ASTM STP 840. (T.E. Allen, ed.) American Society for Testing and Materials, Philadelphia, PA. pp. 69-86.

Byford, D.C., P.R. Laskey, and A. Lewis. 1984. Effect of low temperature and varying energy input on the droplet size distribution of oils treated with dispersants. Proc. Seventh Annual Arctic Marine Oilspill Program Technical Seminar, 12-14 June 1984, Edmonton, Alberta, Canada. Environmental Protection Service, Environment Canada. pp. 208-228.

Canevari, G.P. 1978. Some observations on the mechanism and chemistry aspects of chemical dispersion. In: Chemical Dispersants for the Control of Oil Spills, ASTM STP 659. (L.T. McCarthy, Jr., G.P. Lindblom, and H.F. Walter, eds.) American Society for Testing and Materials, Philadelphia, PA. pp. 5-17.

Canevari, G.P. 1982. The formulation of an effective demulsifier for oil spill emulsions. Marine Pollution Bulletin, 13: 49-54.

Canevari, G.P. 1984. A review of the relationship between the characteristics of spilled oil and dispersant effectiveness. In: Oil Spill Chemical Dispersants: Research Experience and Recommendations, ASTM STP 840. (T.E. Allen, ed.) American Society for Testing and Materials, Philadelphia, PA. pp. 87-93.

Canevari, G.P. 1985. The effect of crude oil composition on dispersant performance. Proc. 1985 Oil Spill Conference, 25-28 Feb. 1985, Los Angeles, CA. American Petroleum Institute, Washington, D.C. pp. 441-444.

Canevari, G.P. 1987. Basic study reveals how different crude oils influence dispersant performance. Proc. 1987 Oil Spill Conference. American Petroleum Institute, Washington, D.C. pp. 293-296.

Canevari, G.P., J. Bock, and M. Robbins. 1989. Improved dispersant based on microemulsion technology. Proc. 1989 Oil Spill Conference, 13-16 Feb. 1989, San Antonio, TX. American Petroleum Institute, Washington, D.C. pp. 317-320.

Canevari, G.P. 1991. The relationship between oil dispersion and herding oil dispersant application. Unpublished report presented at the 1991 International Oil Spill Conference, 4-7 Mar. 1991, San Diego, CA. 5 p.

Capuzzo, J.M. 1987. Biological effects of petroleum hydrocarbons: Assessments from experimental results. In: Long-term Environmental Effects of Offshore Oil and Gas Development. (Boesch, D.F. and N.N. Rabalais, eds.) Elsevier Applied Science, London. pp. 343-410.

CEDRE. (undated) French test for measuring effectiveness of dispersants. CEDRE document no. AFNOR T90-345. Centre de Documentation de Recherche et d'Experimentations sur les Pollutions Accidentelles des Eaux, Plouzane, France.

Chianelli, R.R., T. Aczel, R.E. Bare, G.N. George, M.W. Genowitz, M.J. Grossman, C.E. Haith, F.J. Kaiser, R.R. Lessard, R. Liotta, R.L. Mastracchio, V. Minak-Bernero, R.C. Prince, W.K. Robbins, E.I. Stiefel, J.B. Wilkinson, S.M. Hinton, J.R. Bragg, S.J. McMillen, and R.M. Atlas. 1991. Bioremediation technology development and application to the Alaskan spill. Proc. 1991 International Oil Spill Conference, 4-7 Mar. 1991, San Diego, CA. American Petroleum Institute, Washington, D.C. pp. 549-558.

Clark, R.C., Jr. and D.W. Brown. 1977. Petroleum: Properties and analyses in biotic and abiotic systems. In: Effects of Petroleum on Arctic and Subarctic Marine Environments and Organisms. Volume I. Nature and Fate of Petroleum. (Malins, D.C., ed.) Academic Press, Inc., New York. pp. 1-89.

Clayton, John, Siu-Fai Tsang, Victoria Frank, Paul Marsden, and John Harrington. 1992. Chemical Dispersant Agents: Evaluation of Three Laboratory Procedures for Estimating Performance. Final Report. Submitted to U.S. Environmental Protection Agency, Risk Reduction Engineering Laboratory, Releases Control Branch, 2890 Woodbridge Avenue (MS-104), Edison, New Jersey 08837 by Science Applications International Corporation (SAIC). 184 p.

CONCAWE. 1986. Oil spill dispersant efficiency testing: Review and practical experience. Report prepared by the CONCAWE Oil Spill Clean-up Technology Management Group's Special Task Force on the Review of Dispersant Effectiveness Test Methods (OSCUT/STF-6). Report No. 86/52. CONCAWE, Den Haag, the Netherlands.

Cormack, D.B., W.J. Lynch, and B.D. Dowsett. 1986/87. Evaluation of dispersant effectiveness. Oil & Chemical Pollution, 3: 87-103.

Cox, J.C. and L.A. Schultz. 1981. The containment of oil spilled under rough ice. Proc. 1981 Oil Spill Conference, 2-5 Mar. 1981, Atlanta, GA. American Petroleum Institute, Washington, D.C. pp. 203-208.

Cunningham, J.M., K.A. Sahatjian, C. Meyers, G. Yoshioka, and J.M. Jordan. 1991. Use of dispersants in the United States: Perception or Reality? Proc. 1991 International Oil Spill Conference, 4-7 Mar. 1991, San Diego, CA. American Petroleum Institute, Washington, D.C. pp. 389-393.

Daling, P.S. and R.G. Lichtenthaler. 1986/87. Chemical dispersion of oil. Comparison of the effectiveness results obtained in laboratory and small-scale field tests. Oil & Chemical Pollution, 3: 19-35.

Daling, P.S. 1988. A study of the chemical dispersibility of fresh and weathered crude oils. Proc. Eleventh Arctic Marine Oil Spill Program Technical Seminar. Environment Canada, Ottawa, Ontario. pp. 481-499.

Daling, P.S., P.J. Brandvik, D. Mackay, and O. Johansen. 1990a. Characterization of crude oils for environmental purposes. Proc. Thirteenth Arctic Marine Oil Spill Program Technical Seminar, 6-8 June 1990, Edmonton, Alberta. Environment Canada, Ottawa, Ontario. pp. 119-138.

Daling, P.S., D. Mackay, N. Mackay, and P.J. Brandvik. 1990b. Droplet size distributions in chemical dispersion of oil spills: Towards a mathematical model. Oil & Chemical Pollution, 7: 173-198.

Daling, P.S., P.J. Brandvik, D. Mackay, and O. Johansen. 1990c. Characterization of crude oils for environmental purposes. Oil & Chemical Pollution, 7: 199-224.

Delvigne, G.A.L. 1985. Experiments on natural and chemical dispersion of oil in laboratory and field circumstances. Proc. 1985 Oil Spill Conference, 25-28 Feb. 1985, Los Angeles, CA. American Petroleum Institute, Washington, D.C. pp. 507-514.

Delvigne, G.A.L. 1987. Droplet size distribution of naturally dispersed oil. In: Fate and Effects of Oil in Marine Ecosystems. (Kuiper, J. and W.J. Van den Brink, eds.) Martinus Nijhoff Publishers, Dordrecht, The Netherlands. pp. 29-40.

Delvigne, G.A.L. and C.E. Sweeney. 1988. Natural dispersion of oil. Oil & Chemical Pollution, 4: 281-310.

Delvigne, G.A.L. 1989. Measurements of natural dispersion. In: Oil Dispersants: New Ecological Approaches. ASTM STP 1018. (Flaherty, L.M., ed.) American Society for Testing and Materials, Philadelphia, PA. pp. 194-206.

Dennis, R.W. and B.L. Steelman. 1980. Overland aerial applications tests of oil spill dispersants held at Abbotsford, B.C., March 13-14, 1979. Technical Report, Exxon Research and Engineering Company, Florham Park, N.J.

Desmarquest, J.P., J. Croquette, F. Merlin, C. Bocard, G. Castaing, and C. Gatellier. 1985. Recent advances in dispersant effectiveness evaluation: experimental and field aspects. Proc. 1985 Oil Spill Conference, 25-28 Feb. 1985, Los Angeles, CA. American Petroleum Institute, Washington, D.C. pp. 445-452.

Diaz, A. 1987. A field dispersant effectiveness test. Final Report submitted to Hazardous Waste Engineering Research Laboratory, Office of Research and Development, U.S. Environmental Protection Agency, Cincinnati, OH from Mason & Hanger-Silas Mason Co., Inc., Leonardo, NJ. Report No. EPA/600/2-87/072. 41 p.

EPA (Environmental Protection Agency). 1984. National Oil and Hazardous Substances Pollution Contingency Plan; Final Rule. 40 CFR Part 300, Federal Register, Vol. 49, No. 129, pp. 29192-29207.

Farn, R.J. 1983. Sinking and dispersing oil. In: The Control of Oil Pollution. (J. Wardley-Smith, ed.) Graham and Trotman Publishers, London. pp. 172-197.

Fina. (undated) Oil spill test kit procedures. (Kit marketed by Norpol Environmental Services A/S, P.O. Box 120 N-1364 Hyalstad, Norway)

Fingas, M.F., K.A. Hughes, and M.A. Schweitzer. 1987a. Dispersant testing at the Environmental Emergencies Technology Division. Proc. Tenth Arctic Marine Oilspill Program Technical Seminar, 9-11 June 1987, Edmonton, Alberta, Canada. Conservation and Protection, Environment Canada. pp. 343-356.

Fingas, M.F., M.A. Bobra, and R.K. Velicogna. 1987b. Laboratory studies on the chemical and natural dispersability of oil. Proc. 1987 Oil Spill Conference. American Petroleum Institute, Washington, D.C. pp. 241-246.

Fingas, M.F. 1989. Field measurement of effectiveness: Historical review and examination of analytical methods. In: Oil Spill Dispersants: New Ecological Approaches. ASTM STP1018. (Flaherty, L.M., ed.) American Society for Testing and Materials, Philadelphia, PA. pp. 157-178.

Fingas, M.F., V.M. Dufort, K.A. Hughes, M.A. Bobra, and L.V. Duggan. 1989a. Laboratory studies on oil spill dispersants. In: Oil Spill Dispersants: New Ecological Approaches. ASTM STP1018. (Flaherty, L.M., ed.) American Society for Testing and Materials, Philadelphia, PA. pp. 207-219.

Fingas, M.F., D.L. Munn, B. White, R.G. Stoodley, and I.D. Crerar. 1989b. Laboratory testing of dispersant effectiveness: The importance of oil-to-water ratio and settling time. Proc. 1989 Oil Spill Conference, 13-16 Feb. 1989, San Antonio, TX. American Petroleum Institute, Washington, D.C. pp. 365-373.

Fingas, M.F., M.D. Fruscio, B. White, N.D. Stone, and E. Tennyson. 1989c. Studies on the mechanism of dispersant action: Weathering and selection of alkanes. Proc. Twelfth Arctic Marine Oil Spill Program Technical Seminar. Environment Canada, Ottawa, Ontario. pp. 61-89.

Fingas, M.F., B. Kolokowski, and E.J. Tennyson. 1990. Study of oil spill dispersants effectiveness and physical studies. Proc. Thirteenth Arctic Marine Oil Spill Program Technical Seminar, 6-8 June 1990, Edmonton, Alberta. Environment Canada, Ottawa, Ontario. pp. 265-287.

Fingas, M.F., R. Stoodley, N. Stone, R. Hollins, and I. Bier. 1991a. Testing the effectiveness of spill-treating agents: Laboratory test development and initial results. Proc. 1991 International Oil Spill Conference, 4-7 Mar. 1991, San Diego, CA. American Petroleum Institute, Washington, D.C. pp. 411-414.

Fingas, M., I. Bier, M. Bobra, and S. Callaghan. 1991b. Studies on the physical and chemical behavior of oil and dispersant mixtures. Proc. 1991 International Oil Spill Conference, 4-7 Mar. 1991, San Diego, CA. American Petroleum Institute, Washington, D.C. pp. 419-426.

Franklin, F.L. and R. Lloyd. 1986/87. The relationship between oil droplet size and the toxicity of dispersant/oil mixtures in the standard MAFF 'Sea' test. Oil & Chemical Pollution, 3: 37-52.

Geraci, J.R. and D.J. St. Aubin. 1987. Effects of offshore oil and gas development on marine mammals and turtles. In: Long-term Environmental Effects of Offshore Oil and Gas Development. (Boesch, D.F. and N.N. Rabalais, eds.) Elsevier Applied Science, London. pp. 587-617.

Gill, S.D. 1981. A review of the Suffield aerial dispersant application trials. Spill Technology Newsletter (May-June 1981), 6: 105-120.

Gill, S.D. 1984. An evaluation program for aerially applied dispersants. In: Oil Spill Chemical Dispersants: Research Experience and Recommendations, ASTM STP 840. (T.E. Allen, ed.) American Society for Testing and Materials, Philadelphia, PA. pp. 161-165.

Gill, S.D. and C.W. Ross. 1980. Aerial application of oil spill dispersants. Proc. Third Arctic Marine Oilspill Conference, 3-5 June 1980, Edmonton, Alberta, Canada. Environment Canada. pp. 328-334.

Gillot, A., A. Charlier, and R. van Elmbt. 1986/87. Correlation results between IFP and WSL laboratory tests of dispersants. Oil & Chemical Pollution, 3: 445-453.

Goodman, R.H. and M.R. MacNeill. 1984. The use of remote sensing in the determination of dispersant effectiveness. In: Oil Spill Chemical Dispersants: Research Experience and Recommendations, ASTM STP 840. (T.E. Allen, ed.) American Society for Testing and Materials, Philadelphia, PA. pp. 143-160.

Goodman, R.H. and J.W. Morrison. 1985. A simple remote sensing system for the detection of oil. Proc. 1985 Oil Spill Conference, 25-28 Feb. 1985, Los Angeles, CA. American Petroleum Institute, Washington, D.C. pp. 51-55.

Hunt, G.L., Jr. 1987. Offshore oil development and seabirds: The present status of knowledge and long-term research needs. In: Long-term Environmental Effects of Offshore Oil and Gas Development. (Boesch, D.F. and N.N. Rabalais, eds.) Elsevier Applied Science, London. pp. 539-586.

Jasper, W.L., T.J. Kim, and M.P. Wilson. 1978. Drop size distributions in a treated oil-water system. In: Chemical Dispersants for the Control of Oil Spills, ASTM STP 659. (L.T. McCarthy, Jr., G.P. Lindblom, and H.F. Walter, eds.) American Society for Testing and Materials, Philadelphia, PA. pp. 203-216.

Jordan, R.E. and J.R. Payne. 1980. Fate and Weathering of Petroleum Spills in the Marine Environment. A Literature Review and Synopsis. Ann Arbor Science Publishers, Inc., Ann Arbor, Michigan. 174 p.

Lee, M., F. Martinelli, B. Lynch, and P.R. Morris. 1981. The use of dispersants on viscous fuel oils and water in crude oil emulsions. Proc. 1981 Oil Spill Conference, 2-5 Mar. 1981, Atlanta, GA. American Petroleum Institute, Washington, D.C. pp. 31-35.

Lehtinen, C.M. and Aino-Maija Vesala. 1984. Effectiveness of oil spill dispersants at low salinities and low water temperatures. In: Oil Spill Chemical Dispersants. Research Experience and Recommendations, ASTM STP 840. (T.E. Allen, ed.) American Society for Testing and Materials, Philadelphia, PA. pp. 108-121.

Lewis, A., D.C. Byford, and P.R. Laskey. 1985. The significance of dispersed oil droplet size in determining dispersant effectiveness under various conditions. Proc. 1985 Oil Spill Conference, 25-28 Feb. 1985, Los Angeles, CA. American Petroleum Institute, Washington, D.C. pp. 433-440.

Lichtenthaler, R.G. and P.S. Daling. 1985. Aerial application of dispersants--comparison of slick behavior of chemically treated versus nontreated slicks. Proc. 1985 Oil Spill Conference, 25-28 Feb. 1985, Los Angeles, CA. American Petroleum Institute, Washington, D.C. pp. 471-478.

Lindblom, G.P. and C.D. Barker. 1978. Evaluation of equipment for aerial spraying of oil dispersant chemicals. In: Chemical Dispersants for the Control of Oil Spills, ASTM STP 659. (L.T. McCarthy, Jr., G.P. Lindblom, and H.F. Walter, eds.) American Society for Testing and Materials, Philadelphia, PA. pp. 169-179.

Lindblom, G.P. and B.S. Cashion. 1983. Operational considerations for optimum deposition efficiency in aerial application of (oil spill) dispersants. Proc. 1983 Oil Spill Conference, 28 Feb.-3 Mar. 1983, San Antonio, TX. American Petroleum Institute, Washington, D.C. pp. 53-60.

Mackay, D., J.S. Nadeau, and C. Ng. 1978. A small-scale laboratory dispersant effectiveness test. In: Chemical Dispersants for the Control of Oil Spills, ASTM STP 659. (L.T. McCarthy, Jr., G.P. Lindblom, and H.F. Walter, eds.) American Society for Testing and Materials, Philadelphia, PA. pp. 35-49.

Mackay, D., R. Mascarenhas, K. Hossain, and T. McGee. 1980. The effectiveness of chemical dispersants at low temperatures and the presence of ice. Proc. Third Arctic Marine Oil Spill Program Technical Seminar, 3-5 June 1980, Edmonton, Alberta, Canada. Environmental Protection Service, Environment Canada. pp. 317-327.

Mackay, D. and F. Szeto. 1981. The laboratory determination of dispersant effectiveness method development and results. Proc. 1981 Oil Spill Conference, 2-5 Mar. 1981, Atlanta, GA. American Petroleum Institute, Washington, D.C. pp. 11-17.

Mackay, D., W. Stiver, and P.A. Tebeau. 1981. Testing of crude oils and petroleum products for environmental purposes. Proc. 1981 Oil Spill Conference, 2-5 Mar. 1981, Atlanta, GA. American Petroleum Institute, Washington, D.C. pp. 331-337.

Mackay, D. and K. Hossain. 1982. Interfacial tensions of oil/water/chemical dispersant systems. Can. J. Chem. Eng., 60: 546-550.

Mackay, D., A. Chau, and K. Hossain. 1983a. Effectiveness of chemical dispersants: A discussion of recent progress. Proc. Sixth Arctic Marine Oilspill Program Technical Seminar, 14-16 June 1983, Edmonton, Alberta, Canada. Environmental Protection Service, Environment Canada. pp. 151-153.

Mackay, D. and P.G. Wells. 1983b. Effectiveness, behavior, and toxicity of (oil spill) dispersants. Proc. 1983 Oil Spill Conference, 28 Feb.-3 Mar. 1983, San Antonio, TX. American Petroleum Institute, Washington, D.C. pp. 65-71.

Mackay, D. 1984. Uses and abuses of oil spill models. Proc. Seventh Annual Arctic Marine Oil Spill Program Technical Seminar, 12-14 June 1984, Edmonton, Alberta, Canada. Environmental Protection Service, Environment Canada. pp. 1-17.

Mackay, D., A. Chau, K. Hossain, and M. Bobra. 1984. Measurement and prediction of the effectiveness of oil spill chemical dispersants. In: Oil Spill Chemical Dispersants, Research Experience and Recommendations, ASTM STP 840. (T.E. Allen, ed.) American Society for Testing and Materials, Philadelphia, PA. pp. 38-54.

Mackay, D. 1985. Chemical dispersion, a mechanism and a model. Proc. Eighth Annual Arctic Marine Oilspill Program Technical Seminar, 18-20 June 1985, Edmonton, Alberta, Canada. Environmental Protection Service, Environment Canada. pp. 260-268.

Mackay, D. and A. Chau. 1986/87. The effectiveness of chemical dispersants: A discussion of laboratory and field test results. Oil & Chemical Pollution, 3: 405-415.

Martinelli, F.N. 1984. The status of Warren Spring Laboratory's rolling flask test. In: Oil Spill Chemical Dispersants, Research Experience and Recommendations, ASTM STP 840. (T.E. Allen, ed.) American Society for Testing and Materials, Philadelphia, PA. pp. 55-68.

McAuliffe, C.D., B.L. Steelman, W.R. Leek, D.E. Fitzgerald, J.P. Ray, and C.D. Barker. 1981. The 1979 Southern California dispersant treated research oil spills. Proc. 1981 Oil Spill Conference, 2-5 March 1981, Atlanta, GA. American Petroleum Institute, Washington, D.C. pp. 269-282.

McAuliffe, C.D. 1989. The weathering of volatile hydrocarbons from crude oil slicks on water. Proc. 1989 Oil Spill Conference, 13-16 Feb. 1989, San Antonio, TX. American Petroleum Institute, Washington, D.C. pp. 357-363.

McColl, W.D., M.F. Fingas, R.A.E. McKibbon, and S.M. Till. 1987. CCRS remote sensing of the Beaufort Sea dispersant trials 1986. Proc. Tenth Arctic Marine Oil Spill Program Technical Seminar, 9-11 June 1987, Edmonton, Alberta, Canada. Conservation and Protection, Environment Canada. pp. 291-306.

Meeks, D.G. 1980. Performance of some oil dispersants on oil slicks of varying thickness. Marine Pollution Bulletin, 11: 348-352.

Meeks, D.G. 1981. A view on the laboratory testing and assessment of oil spill dispersant efficiency. Proc. 1981 Oil Spill Conference, 2-5 Mar. 1981, Atlanta, GA. American Petroleum Institute, Washington, D.C. pp. 19-29.

Merlin, F., C. Bocard, and G. Castaing. 1989. Optimization of dispersant application, especially by ship. Proc. 1989 Oil Spill Conference, 13-16 Feb. 1989, San Antonio, TX. American Petroleum Institute, Washington, D.C. pp. 337-342.

Mitchell, J.B.A. and E. Janssen. 1991. The use of additives for smoke reduction from burning pool fires. Proc. Fourteenth Arctic Marine Oil Spill Program Technical Seminar, 12-14 June 1991, Vancouver, British Columbia. Environment Canada, Ottawa, Ontario. pp. 391-397.

NRC (National Research Council). 1985. Oil in the Sea: Inputs, Fates and Effects. National Research Council. National Academy Press, Washington, D.C. 601 p.

NRC (National Research Council). 1989. Using Oil Spill Dispersants on the Sea. Committee on Effectiveness of Oil Spill Dispersants, Marine Board, Commission on Engineering and Technical Systems, National Research Council. National Academy Press, Washington, D.C. 335 p.

Nes, H. 1984. Effectiveness of oil dispersants. Laboratory experiments. PFO-project no. 1410, NTNF, Oslo, Norway.

Nichols, J.A. and H.D. Parker. 1985. Dispersants: comparison of laboratory tests and field trials with practical experience at spills. Proc. 1985 Oil Spill Conference, 25-28 Feb. 1985, Los Angeles, CA. American Petroleum Institute, Washington, D.C. pp. 421-427.

Owen, D. 1991. Bioremediation of marine oil spills: Scientific validity and operational constraints. Proc. Fourteenth Arctic Marine Oil Spill Program Technical Seminar, 12-14 June 1991, Vancouver, British Columbia. Environment Canada, Ottawa, Ontario. pp. 119-130.

Payne, J.R., B.E. Kirstein, G.D. McNabb, Jr., J.L. Lambach, C. de Oliveira, R.E. Jordan, and W. Hom. 1983. Multivariate analysis of petroleum hydrocarbon weathering in the subarctic marine environment. Proc. 1983 Oil Spill Conference, 28 Feb.-3 Mar. 1983, San Antonio, TX. American Petroleum Institute, Washington, D.C. pp. 423-434.

Payne, J.R. and G.D. McNabb, Jr. 1984. Weathering of petroleum in the marine environment. Journal of the Marine Technology Society, 18(3): 24-42.

Payne, J.R., B.E. Kirstein, G.D. McNabb, Jr., J.L. Lambach, R. Redding, R.E. Jordan, W. Hom, C. de Oliveira, G.S. Smith, D.M. Baxter, and R. Geagel. 1984. Multivariate analysis of petroleum weathering in the marine environment---Subarctic. Vol. I and II. In: Environmental Assessment of the Alaskan Continental Shelf, Final Reports of Principal Investigators. NOAA, OCSEAP Final Reports 21 and 22 (1984). U.S. Department of Commerce, Washington, D.C. 690 p.

Payne, J.R. and C.R. Phillips. 1985. Petroleum Spills in the Marine Environment. The Chemistry and Formation of Water-in-Oil Emulsions and Tar Balls. Lewis Publishers, Inc., Chelsea, Michigan, USDA, 148 p.

Payne, J.R., C.R. Phillips, M. Floyd, G. Longmire, J. Fernandez, and L.M. Flaherty. 1985. Estimating dispersant effectiveness under low temperature-low salinity conditions. Proc. 1985 Oil Spill Conference, 25-28 Feb. 1985, Los Angeles, CA. American Petroleum Institute, Washington, D.C. p. 638.

Payne, J.R., J.R. Clayton, Jr., C.R. Phillips, J. Robinson, D. Kennedy, J. Talbot, G. Petrae, J. Michel, T. Ballou, and S. Onstad. 1991a. Dispersant trials using the Pac Baroness, a spill of opportunity. Proc. 1991 International Oil Spill Conference, 4-7 Mar. 1991, San Diego, CA. American Petroleum Institute, Washington, D.C. pp. 427-433.

Payne, J.R., J.R. Clayton, Jr., G.D. McNabb, Jr., and B.E. Kirstein. 1991b. Exxon Valdez oil weathering fate and behavior: Model predictions and field observations. Proc. 1991 International Oil Spill Conference, 4-7 Mar. 1991, San Diego, CA. American Petroleum Institute, Washington, D.C. pp. 641-654.

Payne, J.R., G.D. McNabb, Jr., and J.R. Clayton, Jr. 1991c. Oil-weathering behavior in Arctic environments. Polar Research, 10: 631-662.

Pelletier, E.. 1986/87. Dispersion of crude oil in seawater: The role of synthetic surfactants. Oil & Chemical Pollution, 3: 257-279.

Pritchard, P.H. and C.F. Costa. 1991. EPA's Alaska oil spill bioremediation project. Environmental Science and Technology, 25: 372-379.

Rewick, R.T., J. Gates, and J.H. Smith. 1980. Simple test of dispersant effectiveness based on interfacial tension measurements. Fuel, 59: 263-265.

Rewick, R.T., K.A. Sabo, J. Gates, J.H. Smith, and L.T. McCarthy, Jr. 1981. An evaluation of oil spill dispersant testing requirements. Proc. 1981 Oil Spill Conference, 2-5 Mar. 1981, Atlanta, GA. American Petroleum Institute, Washington, D.C. pp. 5-10.

Rewick, R.T, K.A. Sabo, and J.H. Smith. 1984. The drop-weight interfacial tension method for predicting dispersant performance. In: Oil Spill Chemical Dispersants. Research Experience and Recommendations. ASTM STP 840. (T.E. Allen, ed.) American Society for Testing and Materials, Philadelphia, PA. pp. 94-107.

S.L. Ross Environmental Research Limited. 1989. Rapid test for dispersant effectiveness at oil spill sites. Final Report prepared for the American Petroleum Institute, Washington, D.C. by S.L. Ross Environmental Research Limited, Ottawa, Canada. 28 p.

Shum, J.S. 1988. An improved laboratory dispersant effectiveness test. Final Report submitted to Hazardous Waste Engineering Research Laboratory, Office of Research and Development, U.S. Environmental Protection Agency, Cincinnati, OH from Mason & Hanger-Silas Mason Co., Inc., Leonardo, NJ. Report No. EPA/600/2-88/023. 105 p.

Smedley, J.B. 1981. The assessment of aerial application of oil spill dispersants. Proc. 1981 Oil Spill Conference, 2-5 Mar. 1981, Atlanta, GA. American Petroleum Institute, Washington, D.C. pp. 253-257.

Spies, R.B. 1987. The biological effects of petroleum hydrocarbons in the sea: Assessments from the field and microcosms. In: Long-term Environmental Effects of Offshore Oil and Gas Development. (Boesch, D.F. and N.N. Rabalais, eds.) Elsevier Applied Science, London. pp. 411-467.

Steelman, B.L. 1979. Oil spill dispersant applications: A time and cost analysis. Proc. Oil and Hazardous Material Spills Conf., pp. 84-97.

Tabak, H.H., J.R. Haines, A.D. Venosa, J.A. Glaser, S. Desai, and W. Nisamaneepong. 1991. Enhancement of biodegradation of Alaskan weathered crude oil components by indigenous microbiota with the use of fertilizers and nutrients. Proc. 1991 International Oil Spill Conference, 4-7 Mar. 1991, San Diego, CA. American Petroleum Institute, Washington, D.C. pp. 583-590.

Venosa, A.D., J.R. Haines, W. Nisamaneepong, R. Govind, S. Pradhan, and B. Siddique. 1991. Protocol for testing bioremediation products against weathered Alaskan crude oil. Proc. 1991 International Oil Spill Conference, 4-7 Mar. 1991, San Diego, CA. American Petroleum Institute, Washington, D.C. pp. 563-570.

Wells, P.G. and G.W. Harris. 1979. Dispersing effectiveness of some oil spill dispersants: Tests with the "Mackay Apparatus" and Venezuelan Lago Medio crude oil. Spill Technology Newsletter, 4: 232-241.

Wells, P.G., S. Abernathy, and D. Mackay. 1985. Acute toxicity of solvents and surfactants of dispersants to two planktonic crustaceans. Proc. Eighth Annual Arctic Marine Oilspill Program Technical Seminar, 18-20 June 1985, Edmonton, Alberta, Canada. Environmental Protection Service, Environment Canada. pp. 228-240.

Woodward-Clyde Consultants and SRI International. 1987. Evaluation of oil spill dispersant testing requirements. Final Report submitted to Hazardous Waste Engineering Research Laboratory, Office of Research and Development, U.S. Environmental Protection Agency, Cincinnati, OH from Woodward-Clyde Consultants (Walnut Creek, CA) and SRI International (Menlo Park, CA). Report No. EPA/600/2-87/070. 139 p.

INDEX

APPENDIX A

Preparation Approach for this Book

This book has been prepared to update information for the mechanism of action of chemical agents for dispersing oil slicks, variables affecting dispersant performance, evaluations of laboratory tests for assessing performance, and brief consideration of the relevance of laboratory test results to spills in field situations. The majority of the material for the book has been taken from an EPA-sponsored report on the same subject matter. The approach for obtaining information for the EPA-sponsored report was twofold: (1) thorough searches were conducted in the available scientific literature and (2) direct input was invited and incorporated from an international body of experts on relevant subject matter. Toward the latter end, a workshop of experts was held on 15-16 April 1991 in Edison, NJ to discuss not only the current state-of-the-art but also needed directions for future research for chemical dispersants. In addition to input and discussions from individuals attending the workshop, all invitees were provided with two sequential drafts of the EPA-sponsored report. The experts provided critical review and suggestions on the drafts. Particular appreciation is expressed by the authors to the following individuals for constructive input to one or more versions of the draft reports: Michael Borst, Gerard P. Canevari, Neil Challis, Douglas Cormack, Per S. Daling, Gerard Delvigne, Mervin F. Fingas, Robert Fiocco, Gordon P. Lindblom, Clayton D. McAuliffe, Francois Merlin, Royal Nadeau, Knut H. Riple, Choudhry Sarwar, and Maurice Webb. All comments and suggestions from the preceding individuals were given careful consideration and, wherever possible, appropriate adjustments made to the content of the report.

A total of 45 individuals (32 in the U.S. and 13 from countries outside the U.S.) received copies of the draft reports for review and/or attended the workshop in Edison, NJ. A list of these individuals follows.

Don Aurand
  Director of Env. Health Research
  Marine Spill Response Corporation
  1350 I Street, N.W., Suite 300
  Washington, D.C. 20005

Christian Bocard
  Institut Francais du Petrole
  1 et 4, avenue de Bois-Preau
  92506 Rueil-Malmaison
  FRANCE

Michael Borst
 RCB, RREL
 U.S. EPA (MS-104)
 2890 Woodbridge Avenue
 Edison, NJ 08837-3679

John Brugger
 RCB, RREL
 U.S. EPA (MS-104)
 2890 Woodbridge Avenue
 Edison, NJ 08837-3679

D.C. Byford
 British Petroleum Company plc
 BP Research Centre, Chertsey Road
 Sunbury-on-Thames, Middlesex
 UNITED KINGDOM

Gerard P. Canevari
 G.P. Canevari Associates
 104 Central Avenue
 Cranford, NJ 07016

Neil Challis
 International Tanker Owners Pollution Federation Ltd.
 Staple Hall, 87-90 Houndsditch
 London EC3A 7AX
 UNITED KINGDOM

Carl Chen
 RCB, RREL
 U.S. EPA (MS-104)
 2890 Woodbridge Avenue
 Edison, NJ 08837-3679

John R. Clayton, Jr.
 Science Applications International Corp.
 Loc. 001,C2 - Environmental
 10260 Campus Point Drive
 San Diego, CA 92121

Douglas Cormack
    Warren Spring Laboratory
    Gunnels Wood Road
    Stevenage
    Hertfordshire
    SG1 2BX
    UNITED KINGDOM

John Cunningham
    Chief, Response Standards and Criteria Division
    Emergency Response Division
    U.S. EPA (OS-210)
    401 M Street, SW
    Washington, D.C. 20460

Per S. Daling
    Environmental and Chemistry Group
    IKU Sintef Group
    Continental Shelf and Petroleum Technology Research Institute
    S.P. Andersens vei 15 b
    N-7034 Trondheim
    NORWAY

Gerard A.L. Delvigne
    Delft Hydraulics
    P.O. Box 177
    185 Rotterdam Se Weg
    2600 MH Delft
    THE NETHERLANDS

James Duffy
    ICF
    9300 Lee Highway
    Fairfax, VA 22031-1207

John S. Farlow
    Chief, Env. Response Branch
    RCB, RREL
    U.S. EPA (MS-104)
    2890 Woodbridge Avenue
    Edison, NJ 08837-3679

Mervin F. Fingas
Emergencies Science Division
Environment Canada
River Road Environmental Technology Centre
Ottawa, Ontario K1A OH3
CANADA

Robert Fiocco
Exxon Research & Engineering Co.
P.O. Box 101
Florham Park, NJ 07932

John P. Fraser
23 Hibury Drive
Houston, TX 77024

Jack R. Gould
American Petroleum Institute
1220 L Street N.W.
Washington, D.C. 20005

Norman Goulet
Scientex Corp.
3204 Tower Oaks Boulevard, Suite 400
Rockville, MD 20852

Robert Hiltabrand
U.S. Coast Guard R&D Center
Avery Point
Groton, CT 06340-6096

Chad T. Jafvert
Environmental Research Laboratory
U.S. EPA
College Station Road
Athens, GA 30613

Julie Jordan
Scientex Corp.
3204 Tower Oaks Boulevard, Suite 400
Rockville, MD 20852

Doug Kodama
    Onscene Coordinator
    Response and Prevention Branch
    U.S. EPA (MS-211), Building 209
    2890 Woodbridge Avenue
    Edison, NJ 08837

Thomas P. Kridler
    Science Applications International Corp.
    1710 Goodridge Drive
    McLean, VA 22102

Gordon P. Lindblom
    14351 Carolcrest
    Houston, TX 77079

Tim Lunel
    Warren Springs Laboratory
    Gunnels Wood Road
    Stevenage, Hertfordshire SG1 2BX
    UNITED KINGDOM

Donald Mackay
    Department of Civil Engineering
    University of Toronto
    Toronto, Ontario M5S 1A1
    CANADA

Paul Marsden
    Science Applications International Corp.
    Loc. 001,C2 - Environmental
    10260 Campus Point Drive
    San Diego, CA 92121

Clayton McAuliffe
    1220 Frances Avenue
    Fullerton, CA 92631

Francois Merlin
    CEDRE
    B.P. 72, Pointe du Diable
    29280 Plouzane
    FRANCE

Gary Moore
    Scientex Corp.
    3204 Tower Oaks Boulevard, Suite 400
    Rockville, MD 20852

Royal Nadeau
    Deputy Chief, Env. Response Branch
    Emergency Response Division
    U.S. EPA, Building 18 (MS-101)
    2890 Woodbridge Avenue
    Edison, NJ 08837-3679

Jerry M. Neff
    Arthur D. Little, Inc.
    Marine Science Department
    25 Acorn Park
    Cambridge, MA 02140-2390

J.A. Nichols
    International Tanker Owners Pollution Federation Ltd.
    Staple Hall, 87-90 Houndsditch
    London EC3A 7AX
    UNITED KINGDOM

James R. Payne
    Sound Environmental Services, Inc.
    2236 Rutherford Road, Suite 103
    Carlsbad, CA 92008

Robert T. Rewick
    SRI International
    333 Ravenswood Avenue
    Menlo Park, CA 94025

Knut Riple
    Esso Norge a.s
    P.O. Box 60
    N-4033 Forus
    NORWAY

John Rogers
    U.S. EPA
    Environmental Research Laboratory 150
    College Station Road
    Athens, GA 30613

Karen A. Sahatjian
    Response Standards and Criteria Division
    Emergency Response Division
    U.S. EPA (OS-210)
    401 M Street, SW
    Washington, D.C. 20460

Choudhry Sarwar
    RCB, RREL
    U.S. EPA (MS-106)
    2890 Woodbridge Avenue
    Edison, NJ 08837-3679

Anthony Tafuri
    RCB, RREL
    U.S. EPA (MS-104)
    2890 Woodbridge Avenue
    Edison, NJ 08837-3679

LCDR Peter Tebeau
    U.S. Coast Guard R&D Center
    Avery Point
    Groton, CT 06340-6096

Ed Tennyson
    Minerals Management Service
    Technical Assessment Research
    381 Elden Street
    Herndon, VA 22070-4817

Maurice Webb
    Warren Springs Laboratory
    Gunnels Wood Road
    Stevenage, Hertfordshire SG1 2BX
    UNITED KINGDOM

Printed and bound by CPI Group (UK) Ltd, Croydon, CR0 4YY
23/10/2024
01778230-0014